PRAISE FOR *MINI-FOREST REVOLUTION*

"My late friend and colleague, Professor Akira Miyawaki, wanted nothing more than to repair the forests of the world. He wanted trees in the ground, as do I. This book would make him happy."

—DIANA BERESFORD-KROEGER, author of *To Speak for the Trees*

"We cannot solve problems by succumbing to fear and anger, and yet so much of the climate conversation is powered by the fearful narrative of a dying planet. In *Mini-Forest Revolution,* Hannah Lewis offers a different story—one that is authentic, honest, and powered by love. Her writing provides the inspiration, motivation, and recipe for working with nature rather than against it, for gathering our courage and creating the world we imagine."

—SHUBHENDU SHARMA, founder and director of Afforestt

"Imagine a world where every modest scrap of worn-out dirt or asphalt—think tennis-court-size—can become a cooling, moisture-circulating, air-cleansing, wildlife-nurturing forest within a few years. *Mini-Forest Revolution* shows how ordinary citizens can embrace this trowel-ready solution, and are doing so even under the harshest, sun-bleached conditions."

—JUDITH D. SCHWARTZ, author of *The Reindeer Chronicles*

Mini-Forest Revolution

Mini-Forest Revolution

Using the Miyawaki Method to Rapidly Rewild the World

HANNAH LEWIS
FOREWORD BY PAUL HAWKEN

Chelsea Green Publishing
White River Junction, Vermont
London, UK

Project Manager: Alexander Bullett
Acquiring Editor: Fern Marshall Bradley
Developmental Editor: Natalie Wallace
Copy Editor: Diane Durrett
Proofreader: Laura Jorstad
Indexer: Linda Hallinger
Designer: Melissa Jacobson
Page Layout: Abrah Griggs

Printed in Canada.
First printing May 2022.
10 9 8 7 6 5 4 3 2 1 22 23 24 25 26

Our Commitment to Green Publishing

Chelsea Green sees publishing as a tool for cultural change and ecological stewardship. We strive to align our book manufacturing practices with our editorial mission and to reduce the impact of our business enterprise in the environment. We print our books and catalogs on chlorine-free recycled paper, using vegetable-based inks whenever possible. This book may cost slightly more because it was printed on paper that contains recycled fiber, and we hope you'll agree that it's worth it. *Mini-Forest Revolution* was printed on paper supplied by Marquis that is made of recycled materials and other controlled sources.

Library of Congress Cataloging-in-Publication Data
Names: Lewis, Hannah, 1974- author.
Title: Mini-forest revolution : using the Miyawaki method to rapidly rewild the world / Hannah Lewis.
Other titles: Using the Miyawaki method to rapidly rewild the world
Description: White River Junction, Vermont : Chelsea Green Publishing, 2022. | Includes bibliographical references and index.
Identifiers: LCCN 2022013529 (print) | LCCN 2022013530 (ebook) | ISBN 9781645021278 (paperback) | ISBN 9781645021285 (ebook)
Subjects: LCSH: Miyawaki, Akira, 1928-2021. | Wildlife reintroduction. | Reforestation. | Forest restoration.
Classification: LCC SD409 .L4785 2022 (print) | LCC SD409 (ebook) | DDC 634.9/56—dc23/eng/20220411
LC record available at https://lccn.loc.gov/2022013529
LC ebook record available at https://lccn.loc.gov/2022013530

Chelsea Green Publishing
85 North Main Street, Suite 120
White River Junction, Vermont USA

Somerset House
London, UK

www.chelseagreen.com

In memory of Dr. Akira Miyawaki (1928–2021),
whom I never knew nor had the privilege to speak with,
but whose words, actions, and ever-expanding network
of friends around the world inspire me daily.

CONTENTS

FOREWORD

It is rare that a book describing a climate solution flows like honey and reads like silk. It is equally uncommon to find an action that everyone can do. Hannah Lewis describes a gift to a despairing world, a way to change the Earth in practical, restorative, and substantive ways, a simple act that creates beauty and enchantment: a mini-forest. If planting a mini-forest seems inadequate to the climate task at hand—reversing global warming—bear in mind that forests are thousands of mini-forests under one canopy. The extraordinary forests of the Boreal and the Congo were never planted. Mini forests can be planted by you.

Forests can cover vast areas as in the Amazon and Borneo. A mini-forest can cover vacant islands of land, a highway roundabout, a small portion of a playground in a nursery school. One of the virtues of a Miyawaki mini-forest that Hannah Lewis describes so exquisitely is its potential ubiquity. There are hundreds of millions of prospective homes for mini-forests. And unlike the Amazon, they won't be torched and replaced by soy and cattle.

There are calls to plant a trillion trees in order to "combat" climate change. The goal is to alter the climate crisis as quickly as possible. In these scenarios, trees are objects, sylvan "things" that capture carbon above and below the ground. Tree plantations can be ghost towns, soundless because there are no birds. No birds because there are no insects. No insects because there are no blossoms, nectar, or worms.

Standing armies of trees are the opposite of a forest. Trees are like us. We are social and so are trees. They thrive when they are interacting with a variety of other trees, shrubs, and plants. The Miyawaki Method arose from observations of ancient forests. Dr. Miyawaki saw them as a living interactive entity, not a collection of trees.

Mini-Forest Revolution takes us around the world to explore the extraordinary impact mini-forests are having within diverse types of terrain, climate, and location. Hannah is your tour guide. There may be no single climate solution that has a greater breadth of benefits than mini-forests: water, shade, coolth, pollinators, food, birds, biodiversity, water storage, carbon sinks, clean air. When given the space, mini-forests grow sidewise, not just up. They germinate, they spread, they enlarge. In open land, they are forest "seeds." They generate themselves and all of life.

Much of what we hear about the climate crisis is the rate at which peril is increasing and the lack of sufficient action on all levels of agency. We are surfeited with news of the problems, probabilities, and impacts. It is almost too much to take in. What is missing that lends balance to the news is possibility. Every problem is a solution in disguise, or it would not be a problem. The Miyawaki Method is crucial because it is a possibility that can be implemented by people everywhere, and as Hannah points out, by communities, classrooms, cities, clubs, families, and yes, even countries if they wake up. We do not have to wait for nations, banks, and corporations to act.

In order to fully understand the potential global impact of Miyawaki mini-forests on the climate crisis, we can geek out a bit and do some carbon accounting. There are an estimated 3,300 billion tons of carbon held in terrestrial ecosystems. That is four times more carbon than is in the atmosphere in the form of CO_2. If in the next thirty years, we increase the amount of carbon held in land by 9 percent, we will have brought back to earth all of the carbon dioxide emitted by coal, gas, and oil combustion, deforestation, and extractive agriculture since 1800. That would mean increasing the amount of carbon in our lands by .3 percent per year. We know how to do that employing regenerative agriculture, wetland restoration, managed grazing, mangrove plantings, and reforestation. These practices are at hand and being enacted, but they are impractical for an individual, family, or neighborhood to do. Mini-forests can be done by everyone everywhere. There are five billion acres of degraded land on Earth. A mini-forest will increase the amount of carbon in that land at least ten times, likely much more. If one-fifth of our degraded lands became mini-forests, we would achieve the goal of returning all the carbon emitted into the atmosphere from 1800 until now.

Mini-forests are where we touch life. We explore our place, discover what is native, restore our soil, nurture a small ecosystem that restores life. Watching a mini-forest grow several feet a year, watching it become more complex and beautiful in front our eyes, knowing that it has a direct impact on the biosphere and atmosphere—these relationships feed us. They feed our longing to make a difference, our need to connect to what is regenerative and act. Facts do not change our minds. Actions change our minds. As we get involved with the acts of regeneration Hannah presents here, our sense of self and what is possible transforms. A mini-forest of ideas and hope is born within us as well.

—Paul Hawken

INTRODUCTION

Restoration in Roscoff

On a cloudy Thursday morning in mid-December, a troop of rubber boot–clad grade schoolers and their teacher walked up to a line of crates filled with a variety of small, slender saplings. "Who's ready to plant a few trees into this tiny parcel to make a mini-forest?" I asked them. "And who's brave enough to walk through a bit of mud to do it?" Their bright faces lit up with intrigue as they nodded in approval to both questions. I showed them three groups of plants—some oak trees, which will grow tall, a handful of medium-sized trees, and several smaller trees and shrubs—and instructed each child to take one or two from each group.

They collected their saplings as if they were party favors and then marched purposefully onto the plot to stake out a spot for their trees. Some kids worked efficiently, offering to help their peers once their own trees were planted. Others got distracted by the muddy spots, laughing at the squishy sucking sound their boots made as they walked. The fun was well deserved after a two-week delay caused by days of rain that had drenched the site.

I was tickled at the sense of determination with which the kids dug, checking along the way to see if the holes were big enough for the roots, and then digging deeper if they weren't. Before they left I encouraged them to come back and visit their forest, promising that it would be taller than them next year. The entire front row of kids standing around me inched up on their tiptoes to where my hand hovered just over their heads, indicating a future tree height. "They grow that fast?" called out a boy in the back row.

This was their forest, and I was excited to see what it would become for them: an enchanted woods to explore, a place to discover wildlife, a little beating heart at the edge of town.

I happen to live in France, in a small and ancient coastal village once inhabited by pirates. The sun makes its appearance intermittently most days in this rainy town, and often before the drizzle has stopped completely, making rainbows a regular sighting. I'm from Minneapolis originally. My family and I moved to the northwestern coastal region of Brittany in 2016 because my husband, who is French, was offered a position at a marine research station.

We enrolled our twins in the local public school and soon felt right at home. With the occasional cry of seagulls, salty air, houses older than my native country, and dramatic tides that leave little boats marooned in the harbor twice a day, their masts clanging in the wild winds, the town feels magical to a Minnesota kid like me.

When we moved here I started reading and writing about ecosystem restoration science and practice for a US-based nonprofit organization called Biodiversity for a Livable Climate. Through this work I have been fascinated to learn of the myriad secrets hidden in plain sight about how Earth's systems sustain human life. Scientists have revealed in intricate detail how organisms and species interact to store carbon, increase ecosystem productivity and stability, and regulate water cycles, and also how vegetation cools Earth.

The more I read, though, the more I came to feel I was inhabiting parallel universes. On the one hand was my growing appreciation for the planet's life support systems that have already been pushed to the brink, while on the other hand was the observation that routine daily assaults on these Earth systems are continuing. The physical reality that global emissions must be halved by 2030 from 2010 levels conflicts with the political reality that national pledges to cut emissions have not been backed up by the requisite systemic changes. Global plastic pollution was on track to double by 2030, and yet marine life was already choking on the ubiquitous stuff.[1] Why were our behavior trends going in the wrong direction? The incompatibility between what should happen and what was happening was giving me a bad case of cognitive dissonance.

I wanted to connect my two universes, to open up space for conversations about how fast the climate is changing, why that's happening, what we can do to decelerate the process and build resilience, and what is beyond our control. It's not easy, because the potential collapse of civilization is rather a taboo topic.

That's when I discovered MiniBigForest, an initiative launched in 2018 in the not-too-distant city of Nantes. MiniBigForest helps towns, schools, and private entities plant biodiverse, ecologically functional miniature forests in small urban spaces using the Miyawaki Method. They planted their first forest as a noise and pollution buffer against a proposed road expansion.

I was immediately enthralled by the method, which struck me as both ecologically sound and eminently doable. It seemed like something that could transform the vacant lots, parking lot edges, and patches of lawn, the open spaces that are so familiar and banal in most cityscapes we almost do not see them.

I always loved the exhilarating fantasy in the Talking Heads song "(Nothing But) Flowers"—the image of asphalt becoming ecosystems. But even more exhilarating was the possibility of actually bringing about such change. And with the idea planted, I was ready to take my first steps toward pulling my two universes together.

This book tells the story of how I became fascinated by the Miyawaki Method and set out to learn about the visionary scientist who developed it, the ecological theory it is based on, the how-to steps of the method it prescribes, and the people practicing it around the world.

Chapter 1 introduces the basics of the Miyawaki Method, its underlying ecological concepts, and the global movement that is developing around the method. In chapter 2 we learn about the career and philosophy of Japanese plant ecologist Dr. Akira Miyawaki and how the Miyawaki Method draws on relics of ancient forest in places like India and Japan. Chapter 2 also introduces Maruvan, a Miyawaki Method project in a semiarid area of India that is reestablishing local water cycles by restoring native vegetation. Chapter 3 delves further into the relationship between vegetation and water by exploring mini-forest projects in Maharashtra, India; Buea, Cameroon; and Iran's Qazvin Plain.

In chapter 4 we visit urban centers of ultramodern Europe, where people simply seek to reconnect with nature as if with a long-lost sibling. Paris has integrated the Miyawaki Method into a larger effort to relieve heat stress by boosting urban vegetation. Dozens of cities in the Netherlands are planting mini-forests in collaboration with schools to give children the opportunity to learn about and bond with nature. Chapter 5 explores a few of the not-so-mini-forests that Miyawaki himself planted in collaboration with Japanese multinational industrial companies, cities, and associations in Japan, the United States, China, and elsewhere.

Chapter 6 is set in Beirut, Lebanon, the Yakama Reservation in Washington State, and a working-class borough of East London, where mini-forests are envisioned as a way to help heal communities from trauma or grief stemming from government negligence, colonialism, and poverty, respectively.

Chapter 7 is all about ecosystems: what they are, how they function, the role of humans in relation to the rest of the natural world, and how an understanding of these concepts can guide us in confronting the climate crisis. Chapter 8 offers practical guidance on how to turn a vacant lot into an ecosystem via the Miyawaki Method—from organizing a team to identifying native climax species, preparing the soil, planting, and aftercare. The story concludes in Roscoff, France, where, in February 2021, the town council approved a Miyawaki mini-forest project for a grassy parking lot atop a gentle cliff overlooking the ocean. In December 2021, we planted it.

By sharing stories of mini-forests from around the world, including my own, I hope to illustrate not only a broad international concern over the declining condition of the natural world but also the possibility of acting right where you are to help solve a shared problem that calls for all hands on deck. While the Miyawaki Method is based on a sophisticated understanding of vegetation ecology, it is designed to welcome and encourage public involvement in eco-restoration. It is precisely this combination of veracity and accessibility that makes the Miyawaki Method revolutionary.

CHAPTER ONE

The Miyawaki Method

The logic of Miyawaki's method is to create a permanent canopy of climax tree species directly, *without going through successional stages. This creates a more humid, shady forest microclimate as quickly as possible.*

—Elgene O. Box[1]

Bundled up in a rain jacket and donning a wide-brimmed straw hat, Dr. Akira Miyawaki stood in front of thirty-two potted, twig-sized seedlings, each accompanied by an image of the mature tree it would one day become. "I'm going to give you some of the thirty-two species' names," he explained to the employees of a Toyoda Gosei automotive parts plant in Lebanon, Kentucky. The trainees listened attentively. "It is very difficult to remember all thirty-two, but please try to remember three or four. This is an American beech," Miyawaki said, passing around one of the young trees. He encouraged everyone to feel the seedling, to use all of their five senses to get to know this plant that would become an integral member of the forest community they were about to set in the ground.[2]

"The tree-planting we are going to do at Toyoda Gosei is not for the production of timber or the beautification of the area," Miyawaki declared. "This project is for yourself—you as local community people—to protect your lives from disastrous situations." For some, the statement may have seemed like a stretch. But Miyawaki had witnessed native trees and forests accomplish great feats, from withstanding earthquakes while the built infrastructure around them collapsed, to preventing the spread of fire, to blocking cars from drifting out to sea in the aftermath of a tsunami.

A few days later the newly trained leaders themselves led some 4,000 co-workers and community members in planting 35,000 trees and shrubs along the perimeter of the factory grounds. Within a few years the tiny seedlings they had once held in the palms of their hands would transform into a tall, dense band of forest between the manufacturing plant and surrounding warehouses and farmland. This thin strip of wildlife habitat in an otherwise industrially transformed landscape was meant to cushion the immediate area against extreme weather. It was one of thousands of natural forests planted globally according to the Miyawaki Method.

This technique, which allows for the creation of a mature natural forest in a comparatively small amount of time, is based on a careful calculation of the plant species that are best suited to the local environment. This is exciting in and of itself—a mature forest is a beautiful landscape element; a buffer against extreme heat, polluted air, flooding, and drought; an educational opportunity; and part of an antidote to the global climate crisis. But the Miyawaki Method is also exciting because it can be applied to areas of any size—a fact that has given rise to the term *mini-forest* to describe small dense woods taking root around the world in locations both urban and rural. Imagine turning an area as small as six parking spaces into a forest—it can be done! Calling such a tiny grove a "forest" refers to the natural structure and composition of the vegetation rather than the footprint. Indeed, a true forest is much larger. Yet the implications are radical: With enough dedication, anyone anywhere can involve their community in a process of rewilding depleted land, one small patch at a time.

To See the Forest for the Trees

Miyawaki developed the forest restoration method that bears his name in the 1970s, as Japan's rapid postwar development was showing its downside in the form of pollution and deforestation. The young scientist understood something that is not necessarily obvious: that humans depend on functioning ecosystems for our well-being and survival.

"It is vegetation, especially forests with multiple, complex layers of various trees, that controls a wide range of environmental processes and conditions," Miyawaki wrote in his 2006 book, *The Healing Power of Forests*, coauthored with American ecologist Elgene O. Box.[3]

Forests cover about a quarter of Earth's land surface, yet an estimated 82 percent of forests are degraded to varying extents as a result of industrial logging and other activities. They have lost at least part of their capacity to play their vital, protective roles.[4] Though generally preferable to bare ground, vegetation varies in its ability to provide ecological services. Much of the greenery in cities and suburbs is a combination of trendy flowering plants, isolated trees, and ubiquitous lawn, the latter of which demands mowing and watering to maintain its carpet-like perfection. Crop monocultures dominate rural agricultural landscapes.

The plants we call *weeds* grow in all the in-between spaces, and as Miyawaki came to appreciate, they play a healing role on land, similar to that of a scab on skin. "In nature, land does not want to remain barren," Miyawaki wrote.[5] Yet weeds rarely have the chance to assemble into more complex, leafier communities because of ongoing trampling or clearing; the land they colonize remains sparsely covered. In contrast to single-layer vegetation like weedy patches or manicured lawns, forests are endowed with five to thirty times more green-surface area and are thus "much more effective at providing ecological services."[6]

Planting a forest is not the same as simply planting trees. We plant trees for many reasons: to produce commodities like wood, fruit, oil, or rubber; to decorate and shade yards, streets, and parks; and to block wind, stop erosion, or sequester carbon. Each use determines the species that are chosen and how those trees will be planted. For example, a timber plantation may resemble a natural forest from a distance, but up close, we can see a monotonous grid pattern. The goal is uniform, fast-growing, straight-trunked trees that are easy to access with large harvesting machines. Similarly, if we consider carbon sequestration as a singular goal, we may favor planting only a few fast-growing species to achieve a quick result.

So, what's the problem with planting *trees* rather than planting *forests*?

To put it simply it is the interactions we cannot see that drive the ecological processes we value. The past few decades have seen a rousing surge in research illuminating some of these previously hidden interactions. "A forest is much more than what you see," explains Suzanne Simard, whose pioneering research shows how underground fungal networks connect trees to one another, allowing them to communicate and share nutrients. These webs of exchange enable a forest "to behave as though it's a single organism," with a kind of intelligence.[7] A natural forest is a community of coexisting, interacting organisms—trees, shrubs, moss, fungi, bacteria, insects, animals (including humans acting as equal members of the community)—that rely on one another for food, shelter, and other ingredients of life.

Interspecies interactions fortify the ecosystem as a whole. For example, mycorrhizal fungi—the fungi that form mutually beneficial relationships with plants' root systems—enable plants to transfer carbon into the soil, where it may ultimately be stored for hundreds or thousands of years. These fungi also improve the soil's structure, making it spongy and able to absorb abundant rainwater, some of which infiltrates farther into the ground to refill aquifers. A living soil rich in organic matter is critical to a forest's ability to mitigate flooding and drought. But these vital relationships arise only when plants are allowed to grow and thrive in a natural community. When we plant individual trees or monoculture tree plantations, we miss out on many of the benefits that come from these webs of interdependency.

Just as planting a forest is an improvement on planting a grid of timber trees, planting a forest according to the Miyawaki Method ensures that the forest will be the best fit for its environment—more stable, more resistant to stress, and ultimately more successful.

Most people will never be able to take on a big ecosystem restoration project on the scale that is needed—they will not have the resources or the time. But small groups of people all around the world, in innumerable settings and circumstances, can plant a mini-forest. It is a revolutionary approach to planting trees, and it's taking hold from India to the Netherlands and everywhere in between.

What Is the Miyawaki Method?

Most of us know the term *old-growth forest*, which refers to natural forests that are still mostly free of human disturbance (though not necessarily free of human presence). These forests have reached maturity and beyond—a process that often takes centuries. As a result, they host incredible biodiversity and sustain a complex array of ecosystem functions.

The Miyawaki Method is unique in that it re-creates the conditions for a mature natural forest to arise within *decades* rather than *centuries*. At the heart of the method is the identification of a combination of native plant species best suited to the specific conditions at any given planting site. As we'll see, determining this combination of special plants is not always so straightforward.

More than just the species selection, the Miyawaki Method depends on a small collection of core techniques to ensure the success of each planting. These include improving the site's soil quality and planting the trees densely to mimic a mature natural forest. It's also necessary to lightly maintain the site over the first three years—which can include weeding and watering. Amazingly, though, if the simple guidelines are followed, after that point a Miyawaki-style forest is self-sustaining.

The trees grow quickly (as much as 3 ft [1 m] per year), survive at very high rates (upward of 90 percent), and sequester carbon more readily than single-species plantations. The Miyawaki Method is also special for its emphasis on engaging entire communities in the process of dreaming up and planting a forest. Whether you are three years old or eighty-three, chances are you can place a knee-high seedling into a small hole in the ground. At the very least you can appreciate and cherish the return of quasi-wilderness to a space that was once vacant.

Imagining a Mini-Forest's Potential

The Miyawaki Method calls for planting native species, but not just any natives. In particular, the method involves a careful investigation of what's known as *potential natural vegetation* (PNV). This unusual term refers to the hypothetical ecological potential of a piece of land. Or another way to say it is that potential natural vegetation is "the kind of natural vegetation that could become established if human impacts were completely removed from a site" over an extended period of time.[8] A site's PNV depends on many factors, including current climate conditions, soil, and topography.

How is potential natural vegetation different from the plants we see growing around us in towns and cities? For starters, in almost all developed landscapes, many of the plants are not native to the area, and as such may require maintenance to survive or reproduce.

Given that most of Earth's land surface is significantly altered by urbanization, agriculture, road construction, mining, and the like, it is far from obvious what the original vegetation of any given location would have been. (Original vegetation and potential natural vegetation are not necessarily exactly the same, but they are closely related.) Unraveling this mystery takes curiosity, patience, and persistence. However, thinking about land in terms of its potential natural vegetation is a powerful angle from which to approach ecosystem restoration, because it reveals which species and groups of species are best adapted to a particular environment and therefore more likely to thrive and to support a wider web of wildlife.

To arrive at the potential natural vegetation for a given site, it helps to understand the sequence in which plant communities develop.

Nature's March

If left alone, previously forested land can grow back into mature forest via a process known as ecological succession, wherein the biological components of the ecosystem change over time as larger and longer-lived plant communities colonize the land. As mentioned, this process can take centuries to unfold. A foundational aspect of the Miyawaki Method is that it sidesteps the slow and capricious march of natural succession, instead focusing on those plants that mark the theoretical endpoint of succession.

In nature, the successional process begins when lightweight seeds drift in and germinate on bare ground. Hardy, fast-growing plants—what scientists call pioneer species—such as clover, plantain, and dandelion take advantage of ample sunlight and space. They live short lives, produce a lot of seeds, and shelter the ground in the process. Next to show up are larger perennial herbs and grasses, followed by shrubs and pioneer trees, such as birch, poplar, or pine.

"Each new group of species arrives because the environmental conditions, especially the soil, have been improved; each new species becomes established because it is more shade tolerant than the previous species and can grow up under their existing foliage," Miyawaki wrote.[9] He explains that just when a community of plants appears to be reaching its fullest potential, the seeds of the succeeding community are already germinating in its shade. The species making up each new successional stage tend to be bigger, more shade-tolerant, and longer living than those of the previous stage.

"The plant community and the physical environment continue to interact," Miyawaki explained, "until the final community most appropriate for the environment comes into being, one that cannot be replaced by other plant types. In regions with sufficient precipitation and soil, the final community is a forest."[10]

Theoretically, this final community of plants, known as the climax community, is not easily superseded. Big trees that are considered climax species in their respective environments live for hundreds or thousands of years, forming canopies that shade the interior of the forest, keeping it cool and moist. Climax species shade out pioneer species and dominate the forest.

"In the absence of major environmental change, the climax is normally the strongest form of biological society and is stable in the sense that its dynamic changes are constrained within limits," Miyawaki wrote.[11] Partly on account of the microclimate they create, such ecosystems tend to be more resistant to external conditions, such as heat or drought.

What might climax vegetation look like? There are generally a few different climax communities in a given landscape. Cottonwoods and willows might grow in a river valley while pines and firs populate the nearby mountain flanks. In flatter regions with moderately moist soils, the potential natural vegetation is evergreen or deciduous hardwood species such as laurel, oak, maple, or beech. Miyawaki forests have typically been planted in conditions like this. Not all of Earth's biomes, on the other hand, are dense forest. Places like natural grasslands, desert scrub, and sand dunes, for example, have their own ecological value and should generally not be replaced by forest—Miyawaki Method or otherwise—except perhaps along their riparian corridors.

Making a Mini-Forest: The Basics

Rejuvenating the soil is one of the basics of creating a mini-forest on a degraded site. In fact, it's the critical first step—the goal is to simulate the living soil of a healthy, mature forest. This happens naturally during the stages of ecological succession, but because the Miyawaki Method skips immediately to the climax stage, some preparation is required to compensate. In the absence of a loose soil with plenty of organic matter, trees will struggle to grow properly. In a Miyawaki forest project, the soil is typically recharged by decompacting and amending the site with organic materials (see "Preparing the Soil" on page 140 for more information).

Planting density is another signature of the Miyawaki Method. Conventional wisdom says that plants compete for light, water, and soil nutrients; therefore, plants should have lots of space between them to reduce that competition. But it's not how a Miyawaki forest works. For a Miyawaki forest, the standard planting density is three plants per square meter. This density helps achieve the goal of ecosystem regeneration. After all, in a natural forest, plants are not evenly and widely spaced. Dense planting stimulates mutualistic and competitive interactions among the plants and facilitates connections with soil microorganisms. It also promotes virtuous competition for sunlight, hastening upward growth.

Mulching is a critical component of the Miyawaki Method. After planting, the ground is covered with a thick mulch similar to fallen leaves on a forest floor. Indeed, once the young trees have had a chance to mature, they will contribute leaf mulch to the forest floor naturally. Mulch protects the bare soil from water loss by evaporation, from erosion, and from temperature extremes. Mulch also suppresses weed growth and eventually decomposes into the soil, enriching it.

As they become established over the first few years, the plantings typically need occasional watering and weeding, but after three years the young forest patches are developed enough to shade out weeds and shelter the soil. They are then generally self-sufficient and need no maintenance of any sort—no pruning, no watering, no fertilizing, no pest control—ever.

A New Wave

In 2014, at the age of eighty-six, Miyawaki wrote: "I hope all of the Japanese people plant small saplings with their own hands in order to protect their own lives and those of their loved ones, and to preserve the lush verdure of Japan. I wish to spread the know-how and the results of this ecological reforestation to the whole world."[12]

In fact, the engaged botanist was already well on his way to this goal, having led forest-planting festivals in nearly three thousand locations in eighteen countries over the course of about five decades. Miyawaki was awarded the Blue Planet Prize in 2006 for his global leadership in tackling environmental problems, further spreading his message. Miyawaki's partnership with Japanese industrial companies operating internationally was key to propagating the method widely enough for it be picked up by other groups in what would become the second wave of projects.

This new wave began in earnest in 2014 when the technique got a big internet-assisted booster. A young engineer who had discovered and fallen in love with the Miyawaki Method gave a TED Talk on the forest-planting approach, and the five-minute video became a global hit with well over a million views.[13] Shubhendu Sharma worked at a Toyota plant in Bangalore, India, in 2009 when Miyawaki led a planting event there. Sharma volunteered to help plant and was fascinated to see how a natural forest could be assembled from its component parts—densely planted native shrubs, understory trees, and canopy trees re-creating the multiple layers of a functional forest.

The car company's new grove grew vigorously. Impressed, Sharma dove into further research, immersing himself in the ecological concepts upon which the method rests. He planted a forest in his own backyard and observed its development. Nine months after taking root, the young forest had already attracted ten bird species that he had not previously spotted in his yard.

This experience reshaped Sharma's understanding of forests and opened up a world of possibilities. Sharma had been happy working in automobile manufacturing. It was his dream job. But the Miyawaki Method tickled the same part of his mind that drew him to designing automobile assembly lines.

Sharma left Toyota in 2011 to launch the forest-planting company Afforestt, not because he had always wanted to become an entrepreneur, but because the Miyawaki Method was too good to keep secret. "I am an industrial engineer," Sharma explained in his first major talk as founder of Afforestt. "The goal in my life has always been to make more and more products in less amount of time. So today we are making 100-year-old forests in just 10 years."[14] Starting a company "was a way of helping this methodology get a foot in the door of corporates and other institutions," he told me when we spoke in 2020. "This is something everyone should know."

The engineer wondered if he could put his professional training to work on the diffusion of what he considered an extremely effective system for forest regeneration. To that end, he drew up standard operating procedures (SOPs) for each step in the planting process: identifying the right combination of species, analyzing the soil, mapping out the planting site and workflow, preparing the site, planting, and following up with maintenance. He used these SOPs to train his staff and also made them available as an open-source material. Afforestt has since refined its training resources in the development of an eleven-part video series, which is available for free online (see resources, page 167).

In the first two years, Sharma managed to coax all of two people into signing up to have a mini-forest planted. By the third year, both the concept and the company promoting it were starting to catch on locally. Afforestt has since planted at least 138 forests in ten countries, while a growing number of new organizations and small businesses focused on Miyawaki-style mini-forests have sprouted up in in Europe, the Middle East, Africa, Australia, and North America (see resources, page 167, for a list of organizations).

Miyawaki often partnered with industrial leaders, planting what he termed *environmental protection forests* around factories, steelworks, wastewater treatment facilities, shopping centers, and power plants. He also worked with government ministries, cities, schools, and universities in Japan and abroad. By contrast, the new wave of projects is often community-driven, led by passionate, newly minted forest planters who have emerged from careers ranging from architecture and engineering to business administration and computer programming. These projects are also typically smaller than those Miyawaki took on, starting out as local experiments and often developing into larger programs as community stakeholders gain confidence in the approach.

While the Miyawaki Method is the guiding light for these fledgling organizations, following its somewhat complex prescriptions to a tee is easier said than done. This was especially true before detailed instructions on the method were available in English, which Afforestt had to create for itself.

In its early years, the Afforestt team mixed pioneer species in with climax species in an unwitting divergence from the Miyawaki Method. Fast-growing pioneer plants surpassed the climax species, which tolerate the shade but do not thrive in it. So instead of creating a climax forest directly as the Miyawaki Method is designed to do, mixing pioneer plants in with the climax species created an earlier successional community that slowed the growth of the climax species. Such a forest is biodiverse and ecologically beneficial in its own way, but it is not the Miyawaki Method.

The results spoke for themselves. In some cases, the pioneer species collapsed, creating empty pockets in the forests into which climax species eventually took hold. Once they learned, in 2016, that Miyawaki did *not* plant pioneer species in his forests, the company corrected course; they also started requiring an exhaustive forest survey before planting any forest.

Sharma has taken the technique that Miyawaki spread largely in person, shovel in hand, and carried it forward by sparking the imaginations of a generation coming to grips with the intertwining climate and biodiversity crises. The idea of restoring native ecosystems one tennis court–sized patch at a time is empowering, especially now when it is easy to feel helpless. Sharma's pupils have themselves become the teachers, while their pupils, too, are becoming new teachers, spreading the technique and the ecological principles underlying it far and wide.

Regardless of the context, the people who assemble on a rainy or sunny Saturday to plant a native forest end up with their hands submerged in earthy soil. They finish the day having left a legacy whose growth they will be able to observe throughout their lives. And they are exposed to the tacit message that we are part of nature and as such have a role to play in its healing.

Tonic for an Ailing Planet

In the epilogue to *The Healing Power of Forests*, Miyawaki and Box try to make sense of a world whose life-giving ecosystems and myriad life-forms, including humans, are systematically disregarded and disrespected. Their critique zeroes in on the money-oriented culture that accelerated during the 1980s in the United States and elsewhere, snuffing out the burgeoning social and environmental movements of the previous decades.

The authors observed that the pro-business agenda embraced during this time changed cultural values to the point that aggressiveness, greed, and self-centeredness came to be prized as avenues to "success." The problem with this rejiggered value system is that it diminished people's social values and sense of responsibility toward others.

Over Miyawaki's long career, overconsumption in the Global North continued to the point that we have likely now overshot Earth's ecological limits. Miyawaki and Box pointed to massive species die-offs, accelerating levels of pollution and ecosystem loss, lower standards of living for a growing portion of the world's people, increasing numbers of refugees, and greater reliance on technical fixes as signs of this overshoot.

Yet the authors saw no room for despair, calling it irresponsible to resign oneself to tragic outcomes. "It seems that many people still want to lead fulfilling happy lives. This is significant, because people who want to be happy are much more constructive, responsible, and easier to deal with than people who only want to be gratified."[15]

In addition to choosing to live more frugally, the communal activity of forest planting is a way forward. It is a way to revitalize our common lands, communicate hope, and allow each of us to take responsibility: "It was human activity that turned much of the world's land into an unproductive semi-desert, and it must be human efforts that restores at least some of this area."[16]

It is clear from his words and actions that for all of his passion for plants, Miyawaki, at his core, loved humanity. While our species is but one portion of the planet's exquisitely diverse community of life, Miyawaki insists that each human life is precious.

"Nothing is greater than living," he said in a 2012 interview at the age of eighty-four.[17] "Then what is the reason for living? It is to do something anyone can do anywhere for our tomorrow. I want everyone to do what they can do. I will plant trees for the next thirty years—for myself, my loved ones, for Japan, for the future of the seven billion people living on this planet. Let's plant trees together."

CHAPTER TWO

On the Trail of Ancient Forests

In Shinto, the symbols for the eight hundred myriads of divinities are found in nature, especially in large, ancient trees and dense luxuriant forests, which are feared, respected, and revered.

—Akira Miyawaki and Elgene Box[1]

As a child growing up in a farming village north of Hiroshima, Miyawaki had no particular interest in plants.[2] The ones he knew most about were the weeds he saw farmers battling in their fields. But he had an appetite for education, and his studies at an agricultural high school led him toward biology.

Born in 1928, Miyawaki started college just as World War II was ending. He took his entrance exam in a Tokyo suburb the day after the capital city had been pelted by air raids. As a student, he experienced regular hunger pangs due to a food shortage that marred the relative calm after the war.

Soon after graduating from college in Tokyo, Miyawaki entered Hiroshima University, where he studied weed ecology. Located about a half hour's drive out of town, the university had been damaged but not destroyed by the atomic bomb in 1945. "It was four years since the atom bomb had devastated Hiroshima," Miyawaki wrote in his acceptance speech for the 2006 Blue Planet Prize.[3] "The ceiling of the science department building, which had escaped being burned down, was pitch black and the electric lines still hung down. There were just nine other students beside me, and we studied as hard as we could, at night cooking rice in a camping pot together. I enjoyed it immensely."

Wasting no time, the young weed ecologist went on to pursue a doctorate, studying the morphology, or the shape and structure, of weed roots in relation to moisture availability. Miyawaki recalled that although his weed expertise was not much sought after in Japan at that time, he had been noticed in Germany.

"One day an airmail letter arrived," Miyawaki wrote. "Apparently, my work had caught the eye of Prof[essor] Reinhold Tüxen, who was then director of the Federal Institute for Vegetation Mapping in Germany. 'Weeds are at the point where human activity meets natural vegetation, and are extremely important,' he wrote. 'I am also working in the area; by all means come and join me.'"[4]

Miyawaki went to Germany in 1958 and stayed for two years, slogging daily through damp fields, woodlands, heath, and forests under northern Germany's cool, drizzly skies. He examined the types of plants that were growing together in identifiable communities in each of these different environments. At one point, the young biologist gingerly suggested that the fieldwork was too much in such soggy conditions, to which he said Tüxen replied: "Get out there into the field—there are three billion years of the history of life out there, there is a real-life drama unfolding under our great sun!"[5]

Urging Miyawaki to scrutinize, smell, taste, and feel plants in the field, the German elder drilled into him the importance of fieldwork. It was Tüxen, too, who introduced him to the idea of potential natural vegetation, a concept that became the North Star of Miyawaki's lifelong work.

Near the end of his time in Germany, Miyawaki dreamed of a stand of trees at the village shrine in his hometown. Every

November when he was growing up, a festival took place at the Shinto shrine, where villagers would play music and dance all through the night. He remembered looking up at the strong dark branches of a large tree silhouetted against the night sky and having been struck with an emotional sense of wonder and connection.

Shinto and Buddhism, which have been practiced for some 1,500 years in Japan, had established an important cultural relationship to forests. "In Shinto, the symbols for the eight hundred myriads of divinities are found in nature, especially in large, ancient trees and dense, luxuriant forests, which are feared, respected and revered," Miyawaki wrote in *The Healing Power of Forests*.[6]

"When our ancestors founded a village or town, they cleared the natural forest for development but always built a small shrine to honor the village's guardian god in a forest area which they left intact."[7] These shrine forests are called *chinju-no-mori*, which means "forests where the gods dwell." When shrines were built in places where the forest was already cut, a small forest would be restored around the shrine.

Miyawaki awoke from his dream with a new insight: Maybe that big tree is part of the potential natural vegetation for that area. "The shrine forest has been protected as a sacred place since ancient times," he wrote. "It is forbidden for people to enter, cut trees, or modify them. Isn't it possible that traditional native greenery and forest remain in sacred shrine forests (chinju-no-mori)? If I research sacred shrine forests all over Japan, and not just in my hometown, I might be able to know the potential natural vegetation in Japan," he reasoned. "When I realized that, I felt like I could go home, and I had a little hope."[8]

Though he had learned to identify the plants that were likely to be potential natural vegetation at sites he studied in Germany, Miyawaki wondered if he would be able do the same in Japan, a whole new botanical context. The first place he went upon returning from Germany was the shrine in his hometown and the small forest surrounding it. A couple of species of evergreen oak grew there:

Urajiro-gashi (*Quercus salicina*), or Japanese willowleaf oak; and *Aka-gashi* (*Quercus acuta*), or Japanese evergreen oak. The glossy dark green leaves of these types of oaks do not drop off during the winter, unlike the leaves of deciduous oak species, such as bur oak (*Quercus macrocarpa*) in North America.

At that time, it was generally believed that deciduous oak and red pine forests, because they were so widespread, were the original vegetation of the lowlands throughout the country. Miyawaki, however, was unconvinced by this assumption. "He thought: In warmer areas, we should have an evergreen forest," said Dr. Kazue Fujiwara, a plant ecologist and former student of Miyawaki, when I asked her how Miyawaki verified his hunch that the shrine forest species were potential natural vegetation. Fujiwara explained that evergreen broad-leaved forests grow in places with temperatures that stay above -5°C (23°F) year-round, which is true for the lowlands of southern Japan. In the north, where winter temperatures dip lower, deciduous forest predominates.

Miyawaki surveyed the two evergreen oak species (Urajiro-gashi and Aka-gashi) in the field and found that they were perfectly adapted to the climatic conditions of the south side of the Chugoku Mountains, where his hometown is located, which sits 400 m (1,312 ft.) above sea level. He wondered what species grew in other old shrine forests around the country.

Settling in as a lecturer at Yokohama National University in the Kanto region of central Japan, which encompasses Tokyo and constitutes the archipelago's largest population center, Miyawaki began to survey that region's vegetation, driven by his burning question: What is the potential natural vegetation?

Studying ancient groves offered the botanist important information about the characteristic species of the area's original forests. But he had other clues, as well. In addition to the shrine forests, a few other islands of broad-leaved evergreens could be found around people's homes in the Kanto region. Fujiwara explained that people traditionally planted evergreen oaks along the northern edges of their homes to block the cold north wind. The fact that these trees formed tall, canopied, stable, long-living stands showed that they were well adapted to the climate and soils.

But there was still another clue that evergreen broad-leaved forests were native to the Kanto region. On the floor of the secondary deciduous forests that Miyawaki surveyed were oak and other broad-leaved evergreen saplings, suggesting these forests' progress toward a later successional stage. In short, despite a thorough anthropogenic transformation of the land, traces of the original forest vegetation remained.

Fujiwara had joined Miyawaki's team as an undergraduate in 1966, ultimately succeeding Miyawaki as a professor in the Department of Vegetation Science following Miyawaki's 1993 retirement. Fujiwara's first assignment was in a peat bog area, part of which was disturbed by various human activities, and part of which remained intact. She documented how the species composition had changed in the disturbed part, from which the water had drained as though the peat had been wrung like a sponge. These altered conditions allowed a whole new host of plant species to colonize the area.

Intrigued by the question of why certain plants grow in certain environments and not in others, Fujiwara continued to survey Japan's vegetation from north to south—from industrial sites to sacred groves to mountaintops—along with the rest of Miyawaki's team. "Why does this plant grow here? What kind of species would grow well in this area?" she asked herself. "I'm always thinking about that."

The exhaustive surveying undertaken by Miyawaki, Fujiwara, and dozens of other researchers ultimately resulted in the 1990 completion of the ten-volume, 79 lb. (36 kg) Vegetation of Japan series. These books map out the existing vegetation of the entire archipelago. They also "contain specific proposals compiled by region for the preservation of natural vegetation that is close to the original vegetation of the region," Miyawaki wrote in his Blue Planet Prize speech.[9]

Laying out in words what these botanists would dedicate themselves to in action, the books proposed "the creation of disaster prevention and nature conservation forests native to the area, forests that nurture river sources, urban forests, forests in industrial areas, and forests to protect the environment of roads and traffic facilities."[10] In other words, the books contain a prescription for healing Japan's ecosystems.

Miyawaki later wrote *The Healing Power of Forests* in collaboration with Elgene O. Box, a now-retired professor of vegetation science and ecology at the University of Georgia. Box had worked with Miyawaki and Fujiwara due to his interest in the comparability between the natural landscapes of East Asia and eastern North America. Located at roughly the same latitude and influenced by the proximity of the ocean, these two regions have similar climates.

In *The Healing Power of Forests*, Miyawaki and Box set the context for ecosystem restoration by laying out what has been lost. They write that there is almost no virgin forest "in the strict sense" left in Asia, and that by the end of the Roman Empire, "the original forests of the Mediterranean coastal regions, forests of evergreen oak . . . were already largely gone."[11] While development and deforestation happened over many centuries in much of Europe and Asia, they note, it took settlers only about one century to shrink the old-growth forests of North America to a fifth of their former area.

French explorer Alexis de Tocqueville offered a glimpse of this rapid change in his account of trekking through the forest toward Saginaw, Michigan, in 1831. His aim was simply to taste the vast wilderness, long gone from Western Europe. Failing to convince any white Americans of the simple reason for his expedition, he and his traveling companion were compelled to introduce themselves as land prospectors in order to get help with directions. It was incomprehensible to the direction-givers that the wilderness could be something to be savored.

Arriving at his destination after a few days' hike, Tocqueville was awed by the deeply soothing beauty of the vast, intact forest as seen from his perch in a canoe on the Saginaw River. What made Tocqueville's experience all the more arresting was his awareness of the forest-clearing frenzy of his day:

> All that one feels in passing through these flowery wildernesses where everything, as in Milton's Paradise, is ready to receive man is a quiet admiration, a gentle melancholy sense, and a vague distaste for civilized life, a sort of primitive instinct that makes one think with sadness that soon this delightful solitude will have changed its looks. In fact, already the white race is advancing across the forest that surrounds it, and in but a few years the Europeans will have cut the trees that are now reflected in the limpid waters of the lake and forced the animals that live on its banks to retreat into new wildernesses.[12]

Indeed, Saginaw became a timber town not long after Tocqueville's writing, then a manufacturing hub, eventually falling into a postindustrial slump in the twentieth century. The landscapes where most American cities are located are so dramatically changed from what they were a few hundred years ago that it is hard to imagine them as anything else, and certainly not as forest. To most of us, the landscape into which we were born looks perfectly normal, as though it has always looked more or less that way.

Sacred Forests

Just as Japan's chinju-no-mori offer a glimpse into the past, India has ancient and sacred forests that were left standing when towns were settled. The tradition of sparing small forest groves is likely to have started many millennia ago, becoming more significant when agriculture took hold and forests were cleared more readily.[13] Sacred forests were originally thought to house pagan gods, which have since been absorbed into modern religions.[14]

I visited a sacred forest just outside the small town of Meenangadi in the southwestern Indian state of Kerala in January of 2020. My guide was Raju Sankaran, an ornithologist and program leader for the Kerala environmental nonprofit Thanal. We went there to see how a 1.6 ha (4 ac.) reforestation project involving eighty-nine indigenous species of trees and shrubs had restored stream flow in the valley below, where a shrine stood in a small clearing.[15]

By design, the tiny creek flowing from the forested hill runs right through the shrine, the stream directly linking forest and prayer. A few decades ago, Sankaran recounted, the stream started running dry for part of the year. This change had come on the heels of the temple overseers' decision to cut and sell wood from part of the forest when they were in need of funds.

The unintended effect on stream flow troubled the temple community, so they went to the mayor in search of a solution. Town officials connected the temple with Kerala's Forests and Wildlife Department, whose stated objectives include conserving and expanding natural forests, protecting water and biodiversity, and preserving sacred groves.[16] The department agreed to replant the clear-cut section of the sacred forest.

My visit took place five years after the area had been replanted. While the mature forest below was tall and dense, the replanted area resembled a fruit orchard—the trees were still smallish, letting plenty of sunlight onto the red soil, which was covered in sparse grasses and herbs. A couple of temple members on site that day told the story's happy ending: The stream now flows year-round once again, and the forest as a whole seems more alive. Sankaran pointed to a paradise flycatcher darting in and out from the forest edge, drinking and bathing a few meters upstream of the temple. His trained ear picked out another couple dozen bird species calling from the surrounding wooded hills.

"As a result of the long-term conservation of forest patches by communities who consider them to be sacred, relic patches of once extensive forest have been preserved," states a 2010 study of the conservation value of India's sacred forests.[17] Thus, like Miyawaki, Shubhendu Sharma and his Afforestt team survey such groves whenever they can be found near a future planting site.

When beginning to research a new planting site, the Afforestt crew finds local people who can give them a tour of the most natural local forest, whether sacred or not, to identify the species present, not all of which are necessarily native. During one of our conversations, Sharma screen-shared pictures from a forest visit he did with local experts in the northern Himalayas. He explained that locals tend to mention native trees in connection with religious anecdotes. "That kind of validates the history of the origin of that species," he said, because religious texts date back centuries or millennia, documenting people's observations of particular places.

Afforestt staff then consult a variety of secondary texts—scientific, historic, religious, and cultural—to cross-check the local origin of the trees they observed. "All our epics, all our folklore, all our religious texts have been scripted around the forest, or in a forest, and its specific trees," Sharma said. The *Ramayana*, for example, is an epic Vedic Sanskrit tale estimated to have been written some 2,600 years ago. "If you read two pages," Sharma attested, "the action would have happened in just four lines. The remaining text of two pages would be describing the environment of that time—what the protagonist is looking at, what trees are around him. So, it's very easy for one to recreate that environment. The details are exceptional. The main character, Lord Rama, spent 14 years in the forest in the story, so those 14 years you would be reading just about forest—the forest people, the trees, the animals."

Familiarity with texts such as this helps the team recognize native species wherever they go.

A Forest in the Desert

When Afforestt's Gaurav Gurjar was trying to identify local species to plant in a patch of degraded, semiarid land in central Rajasthan, India, he studied old local paintings and poetry. Immersing himself in this particular restoration project, Gurjar actually lives on site on a 30 ac. (12 ha) parcel of desert sandwiched between the 308 mi. (496 km) Luni River and one of its tributaries, the Meethadi. He moved there in 2018 with his life and work partner Varsha Gurjar, Afforestt's training and communication coordinator. The land was offered to Afforestt by a person who runs a school and animal shelter on an adjacent site. His intention was to allow the group to test the Miyawaki Method in an extreme environment and to start a native desert tree nursery. Once established, the project, dubbed Maruvan (meaning "desert forest"), branched off from Afforestt to become an independent nonprofit.

When the Gurjars arrived, the sandy ground was almost completely bare, and thus quite literally a blank slate for experimenting with restoration in a dry environment. "Whenever we do a project for a client, we don't have much scope for experimentation," Gurjar explained to me during a video call. A young man with round glasses, Gurjar seems to be in his element at Maruvan, where he can study ecosystems processes every day. "We only implement those ways of which we are sure, our standard ways of reforestation, and we can't fail on these projects. So Maruvan was started with the view that we are free to fail. For that, we chose one of the most difficult lands. We wanted to grow forests in arid landscapes."

The average annual rainfall in the area is 300 mm (12 in.), but it does not necessarily rain every year, and every twenty to thirty years flash floods wash the landscape down to its bedrock in spots, stripping the land of young and sparse vegetation. Summer temperatures reach highs of 50°C (122°F), while lows in the winter are just above freezing. When I asked Gurjar about trees, he responded by talking about water. Noting that the local hydrological cycle had been broken due to land degradation, Gurjar said his first task was to create a water-harvesting structure.

"You have to start somewhere," Gurjar explained. "If you start with the forest, then you have to externally bring water and then plant it. And once the forest develops, it creates a soil structure in such a way that whatever rain is falling, the soil is able to preserve that moisture in the root zone. But if you don't have water, what will you start with? We started here with water."

Gurjar studied traditional well and pond designs, which desert communities throughout Rajasthan have used for thousands of years to collect and stock water to make it through long rainless periods. Designs vary according to local geology and ingenuity.

After some experimentation, the Maruvan team settled on an open well 6.7 m (22 ft.) deep so as to avoid tapping into the groundwater, which is saline and unsuitable for consumption by plants or people. Instead, the *sajja ka kuva* well, as it is known in local language, harvests subsoil water seeping through horizontal veins in the ground; this water enters through the spaces between the bricks that form the sides of the well and are not sealed with mortar. Subsoil water is in turn fed by a .4 ha (1 ac.) pond the team has dug, along with a channel to connect it to the seasonal river. The pond and well filled up the following monsoon season, providing about eight months of water with which the first Miyawaki-style forest patches could be planted.

Maruvan later expanded the network of ponds, channels, and open wells, anticipating that it would provide enough water to last through two or three years without rain. The water from the ponds seeps beyond the banks underground, eventually reaching the well.

"We'll be studying horizontal movement of water—how far the moisture travels horizontally in the soil," Gurjar explained. "On those horizontal zones, we'll be planting forest. Our target is that we should not need irrigation, because in the desert nobody irrigates the soil—that's not how the trees naturally grow. But to achieve that natural system, first we have to have the natural water systems that can sustain those forests."

Gurjar's Afforestt staff title of Jungle Tree Expert is not for nothing, given his clear knowledge of plants, although he could also be called Drylands Ecosystem Expert. A year before moving on site, he started observing what was happening on the land and what species were present. "I used to sit here for six or seven hours in extreme heat—I used to sit under a tree and just observe what the ants were doing, what the dung beetles were doing, what the hoverflies or dragonflies were doing."

Close-up observations like these have led Gurjar to important discoveries, such as how to get an important local tree to germinate. When Gurjar had first tried to germinate *jaal* tree (*Salvadora persica*) seeds collected directly from the plant, it did not work. "Just out of chance," he explained, "I identified a spot where a lot of bulbul and house sparrow birds sit, and beneath them I found a lot of these seeds were germinating. When I observed their excreta, I saw that there were seeds. Then we started collecting those seeds, and when we germinated them, within one or two days they sprouted up."

Gurjar was able to recognize another bird-plant relationship thanks to the poetry written decades earlier by a local Marwari poet (Marwari refers to the people of the Jodhpur region of southwest Rajasthan). Among 700 verses describing local plants, animals, and their interactions is a verse about the *daabh* grass (*Desmostachya bipinnata*) nests constructed by weaver birds. Neither species was to be found at Maruvan, however, until Gurjar decided to remove invasive *Prosopis juliflora* plants, which is a type of mesquite native in the Americas.

"When we removed the non-native species, that grass automatically returned," Gurjar said. "And as soon as that grass came, weaver birds came flocking to our land. Right now, in my balcony where I'm sitting, I have fifteen to twenty nests that are being created in front of me by the weaver birds. And then the villagers all came saying, 'Okay, this bird has come back after having been missing for forty years on this particular land.'"

Maruvan's first Miyawaki mini-forest patch included forty-four species determined by studying the Luni River basin environment throughout the region, the botanical poetry mentioned above, and local art. Art displayed in a temple a couple of miles from Maruvan depicts an event that occurred in 1730, wherein the king's men killed 363 people as they defended large *khejri* trees (*Prosopis cineraria*) that the king desired cut for the construction of a palace.

"Paintings in the temple are of these large trees, where the women are hugging the trees," Gurjar explained. Similarly, he added, paintings in a nearby fort show the king hunting in the area. Dense vegetation, huge trees, birds, and leopards depicted in these scenes are painted in such detail that individual species can be identified.

These old works of art suggest how the vegetation of this area might have looked before the land became degraded. Gurjar explained that over time, the harvesting of large, slow-growing trees was so great that a tipping point was eventually reached. The hundreds of saplings that would have grown up under the shelter of a single tree canopy could not endure the otherwise extreme conditions, and the cycle of renewal was broken. If tree cutting started the desertification process, then overgrazing, sand mining, and non-native species introductions finished it off.

Even in a healthy state, the vegetation of a semiarid environment differs from that of humid forests elsewhere in India. Gurjar has observed that desert vegetation grows in dense, scattered clusters of grasses, shrubs, and vines around one or two trees. The shrubs preserve enough moisture for the trees to grow. Modeling this structure, the Gurjars and volunteers from the local village planted the forests in circular clusters of four to five seedlings per square meter, each circle spaced apart by a foot or two.

With each successive planting and ongoing observation, Gurjar has whittled down the species list to the most hyper-local adapted plants. "These ecosystems are so complex that just 100 m. [328 ft.] from where I'm sitting, the vegetation or guild could be different," he said. The result is that the second and third groves, planted with thirty-five and twenty-five species, respectively, have demanded less water than the first, which presumably includes species adapted to slightly different conditions elsewhere in the region.

By connecting the land to the river and favoring the best adapted species, Gurjar seems to be on his way to fixing a broken local water cycle.

CHAPTER THREE

Woods and Water

The multiple water and climate-related services provided by forests—precipitation recycling, cooling, water purification, infiltration and groundwater recharge, not to mention their multiple traditional benefits (food, fuel and fiber)—represent powerful adaptation opportunities that can significantly reduce human vulnerability.

—David Ellison et al.[1]

In India, Sharma told me, "People are asking for forests that will give them fruit, shade, clean air, and privacy, as well as bring back biodiversity." While recharging water tables is generally not among the reasons why clients request Afforestt's services, Sharma has seen young forests have that effect. This is important in light of persistent water stress in many parts of India. On account of groundwater depletion in the southern cities of Bangalore and Chennai, for example, people have had to depend on trucks hauling water into the city when taps run dry. Large swaths of land in central and northern India have become desertified, which is what happens when naturally dry land is subject to degradation.[2]

Maharashtra is a large, populous state in western-central India, home to the city of Mumbai. Thanks to the usual causes, especially urbanization and industrial development, this state also contains some of the country's most degraded land.[3] The nonprofit Maharogi Sewa Samiti, Warora (MSS), which cares for socially marginalized people, including those affected by leprosy, is based in a flat, rural, largely treeless stretch of eastern Maharashtra. The leaders of the organization were concerned about ongoing deforestation in the region, which was causing tigers and leopards to move in search of new territory, resulting in increased confrontation and conflict between these animals and local residents. MSS staff wanted to reestablish habitat for indigenous plants, animals, and microorganisms, thereby "completing the circle of life in [these] absolutely barren and devastated lands."[4]

In May 2019, the nonprofit hired Afforestt to implement the Miyawaki Method on a parcel of land slightly smaller than a football field—just under .4 ha (1 ac.) of the organization's 27 ha (67 ac.) property. This land, which had been used to produce firewood from a non-native tree species, was otherwise patchily covered in grass, scrub, and small clumps of trees.

The Miyawaki planting was part of what is to become MSS's biodiversity park, meant to provide wildlife habitat; supply fruit, honey, and native seed stock; and offer ecological learning experiences for local and regional students. Two other forest plots were planted around the same time as the Miyawaki planting—one a slight variation on the Miyawaki Method, and the other a more conventional less-dense planting. All three plots are populated with a wide range of native species, for a total of some 45,000 trees on about 1.2 ha (3 ac.).

Within a year of planting, the nonprofit staff and visiting students started spotting porcupines, peacocks, snakes, blue-bills, dragonflies, and a sloth bear in the young forest patches. At least one panther has discovered the welcoming new territory, as evidenced by paw prints on the ground.

In addition to the wildlife, the water table appears to have risen. As shown in a report to the United Nations Development Programme about this forest-planting project, an angular human-made lake at the southern end of the property is visibly low in a photo taken in February 2019, prior to the planting, with exposed sandbars stretching out from the shores. A photo taken in March 2020 shows the lake much fuller: Water reaches all the way to the shrub-bordered shoreline.[5] Sharma says measurements taken by the nonprofit showed that a well near the planting site also rose significantly in the first year after the forests were planted.

This is only logical, Sharma explained, because forests create spongy soils that absorb rainwater and cool the land through shading and transpiration, the process by which plants release water vapor. Forests also release pollen grains and plant and mushroom spores into the air. In clouds, water droplets form around these biogenic aerosols, helping to generate more rain, especially for downwind areas.

"Do you see this thermometer and humidity meter?" Sharma asked me, sharing pictures on his computer screen of two digital thermometers. One sat on the leaf-littered floor of the then six-month-old Miyawaki mini-forest in Maharashtra, and the other rested on an adjacent dirt road. Both readings were taken at 10:15 a.m. The temperature inside the forest was 14°C (25°F) cooler, and the humidity 43 percent higher. "The ground outside the forest is absolutely hard, dry soil," Sharma said, "versus inside the forest, where it is moist, soft, and full of life."

"When the soil is alive, earthworms move around and bring the mineral-rich soil at the lower levels to the upper levels," Sharma continued, explaining that earthworms also help form tunnel-like spaces through which water can flow. "So, when it rains, the rainwater just travels through these tunnels and directly feeds the groundwater table. And when that happens, your soil erosion stops because water is not striking hard surface and running off, but trickling through and going to the groundwater table."

Similarly, he said, "sunlight falling on bare ground heats it. Yet, the moment you plant a forest, the sunlight gets absorbed by leaves for photosynthesis. All that hot sunlight is being consumed and trapped in the form of fruits, in the form of biomass, in the form of sugars, tubers, roots, et cetera. When that happens, the ecosystem creates an environment where the trees will conserve the groundwater and also release some moisture into the air continuously. So, with time, your groundwater table is only going to increase and your rivers are going to turn perennial."

Planting the Plain

Nearly 2,000 miles (3,219 km) and three countries to the west of Maruvan (see "A Forest in the Desert," page 30), one of Afforestt's first international projects was in a similarly dry area of northern Iran. Situated between Iraq and Afghanistan, Iran's climate ranges from hot desert to cold mountains to mild Mediterranean to temperate forest along the shores of the Caspian Sea. On account of such topographical and climatic variety, Iran is quite biodiverse. Even in the relatively mild northeast, however, annual rainfall is limited and crops often depend on irrigation—92 percent of the country's water consumption is attributed to agriculture.[6]

Sharma's Iranian client was Bahram Torkaman, a friendly gentleman with a handsome white beard who runs a small dairy farm in the semiarid Qazvin Plain, at the southern foot of the Alborz Mountains about an hour west of Tehran. He sells to a nearby milk plant. Enclosed by a perimeter wall, Torkaman's .8 ha (2 ac.) farm sits in a wide-open landscape of continuous unvegetated ground stretching to mountains on the horizon. This is cropland for input-dependent wheat, barley, and hay, greening up occasionally during the growing season thanks to irrigation. Like Gurjar, Torkaman began by talking about water when I spoke with him by phone.

"The most important thing in Iran is water," Torkaman said. He is disappointed with the cropping practices he sees around him. "The agricultural system is copied from Western agriculture—they increase production with chemicals, which makes the land very thirsty. So it needs more water. Water shortage is felt strongly." It was this observation that spurred him to contact Afforestt about the Miyawaki Method. "Iran is a dry country, so it needs such a system—forests that can grow without being watered."

When Afforestt staff visited in 2016 to help Torkaman plant a mini-forest, the intrepid team was surrounded by doubters. How could a forest grow on land that cannot even sustain small, short-lived crops without irrigation? Torkaman explained that in most local people's experience, a plant dies within two weeks if it isn't watered. Sharma recalled how one of Torkaman's dairy employees, tasked with the legwork of preparing the ground with composted cow manure, planting, and mulching, all per Afforestt's instructions, balked at every step, convinced the project would fail.

The University of Tehran faculty member who consulted Torkaman on native forest species was also incredulous. "I will resign if your forest succeeds," wagered the professor, doubting that trees could survive the region's dry conditions. Still, he helped Torkaman compile a list of about 100 woody species whose native range encompasses the Qazvin area. Torkaman also talked to local people who told him what trees they had observed growing well during their lifetimes.

"The first mission for each of us when doing a Miyawaki project is to find out what species have been here in the past, many of which have gone away," Torkaman related. "But even if we find the species name, there is not always a local supply for it." Indeed, of the identified species, less than half were available. None were available through commercial nurseries, but only from a nursery associated with the university.

Torkaman's team planted forty-seven species, including fig (*Ficus carica*), mahaleb cherry (*Cerasus mahaleb*), and barberry (*Berberis vulgaris*), along with maple (*Acer platenoides*), ash (*Fraxinus excelsior*), oak (*Quercus ilex*), and other species whose ranges extend well north into Europe. During the first three years, Torkaman watered each of his 1,200 plants no more than .5 L (1 pt.) per day, and only on hot or dry days. He watered only three times over the fourth and fifth years combined, and yet his growing young forest strip remained a shock of green in the tan landscape. Neighboring farmers were surprised by his results and became curious to learn more.

Anxious to spread the idea in Iran, Torkaman and his son launched a small Miyawaki forest planting company called Masaleh Sabz Apadana (Apadana Green Materials). Referencing a hall with 100 pillars in Persepolis, the capital of ancient Persia, and the color green, the company's name poetically evokes the image of a forest. The father-son team, for whom the work is a labor of love rather than an income source, have since planted a couple of tiny native forests in the Persian Gulf region.

When It Rains, It Absorbs

Limbi Blessing Tata is a focused, energetic person with a vision as big as her heart. "Restoring forests to me is about more than fighting against climate change. It's about home. We grew up in forests and know forests as our source for everything... fruit, nuts, mushrooms, spices, firewood, medicine," Limbi Blessing Tata said when I asked her about the Miyawaki mini-forest patches she is planting in Buea, Cameroon. "Being able to remake forests, it's just something that gives me a lot of joy. It makes me feel at home."

Buea is an agglomerated municipality in southwestern Cameroon, on the eastern slopes of Mount Cameroon near the coast of the Gulf of Guinea. You can locate Buea in your mental map by picturing the place where Africa's western lobe turns sharply south, where West Africa becomes Central Africa. This hot and humid region, hovering barely north of the equator, is within the Congo Basin rainforest, which stretches from here to the western edge of Uganda in the east. The volume of water in the Congo River, which empties into the Atlantic south of Cameroon, is topped only by that of the Amazon River. Likewise, the Central African rainforest is the world's second largest after the Amazon.

Buea gets plenty of rain—an annual average of 4,000 mm (157 in.)—and also receives water from streams that originate higher up the mountain. In fact, the area around Mount Cameroon is the wettest part of the country. Yet Buea is struggling through an escalating water crisis with roots in a 1980s economic crisis. At that time, Cameroon's economy tumbled when global commodity prices fell.

The country's second largest employer was a state-owned agribusiness called the Cameroon Development Corporation (CDC), which operated on the lower slopes of Mount Cameroon due to its rich volcanic soils. With government support, the company's banana, rubber, and palm oil plantations continued to operate through the crisis while unemployment elsewhere around the country was mounting. This situation drew an influx of workers to Buea.

As people arrived, trees were cut to build houses and construct roads, Tata explained. The opening of the University of Buea in 1993 drew students from around the region, further urbanizing the area. The local population grew from 11,000 in 1970 to 90,000 and rising in 2013.[7] Supply could not keep up with demand, and by 2013 the local water utility's capacity was sufficient for only about half the Buea population.[8] Water had to be rationed, giving people access a few days per week according to a set schedule, Tata said.

While aging infrastructure and a ballooning population contributed to the problem, Tata points to deforestation as a major cause of the dropping water table. In addition to the houses, roads, and large CDC plantations, thousands of smallholder farms in the area grow corn, plantain, cocoyam, vegetables, cocoa, and fruit, Tata explained, because the vast majority of locals depend on these products to make a living.

Bare, paved, and cultivated surfaces increasingly occupy what was once forest. In that sense, Buea's modernization story is not much different than that of thousands of places around the world. From California to Texas, China to France, South Africa to India, to name just a few examples, drought and flooding are at least partly attributable to the loss of forests, grasslands, and wetlands capable of absorbing heavy precipitation and recharging aquifers.

Tata explained that trees have been cut down around all of Buea's water catchments, which are natural springs equipped with collection and distribution systems to supply water to surrounding neighborhoods. In these spots, the water table has historically been so high that water has gurgled out of the ground and then been pumped into large tanks. Water is piped into homes from these tanks in the wealthier areas, while people in more modest neighborhoods collect water from taps on the tanks.

In the past, this system supplied water all the way through the dry season. But in recent years, the catchments were running out of water for part of the year, and so people have resorted to digging wells, which are expensive and still largely insufficient. "The water table keeps going down and we try to go deeper," Tata said. Those without a well have to walk more than a mile to fetch fresh water from the mountains. "If you're lucky to be in a neighborhood where someone is rich and kind," Tata said, "they will dig a well, but the majority of people have to go to the river. However, the accessible rivers and streams are polluted, so they have to go into the forest for clean water."

When Tata founded her organization, Ecological Balance, in 2016, it was not this local water crisis that spurred her into action. Rather, it was her desire to help girls access the educational opportunities they craved. Many Cameroonian girls do not have the chance that Tata did to advance beyond primary school.

Because she was an only child, Tata's parents invested their resources in her education and sent her to secondary school, where she was the only girl in her class. "My case was really special, but at the time I didn't realize how special it was. When you are privileged, though, you put your head into it. I got scholarships and ended up with a postgraduate certificate," she said. She went on to study botany and conservation at the University of Buea.

Tata had met girls as young as eleven doing domestic work to finance their own education. She wondered what she could do to support the average girl in her community who wanted to study. She thought about scholarships, but she realized a cash resource would not stretch very far.

Then she remembered the forest. "Nature is our only resource, our only asset" she observed, while asking herself: "But how do we get a livelihood out of the forest?" When she launched Ecological Balance with the intention of using the environment as a tool for social change, the core programming goal was to train women to sustainably harvest and lightly process forest products.

Participating women learn to extract essential oils from seeds; dyes from seeds, leaves, and barks; and fragrances from flowers. They also grind seeds into spice powders and process medicinal plants into tinctures. The finished products sell for up to ten times the value of their raw counterparts, and the income supports these women's daughters' educational pursuits.

In 2018 the botanist-conservationist had the chance to further advance her own education, this time focusing on her capacity to grow into the leadership role she had created. She enrolled in a yearlong program at Kanthari Institute for Social Change in Kerala, India, hoping to expand her networks, which she saw as essential to NGO survival and effectiveness. Working with other budding leaders from around the world, she learned how to run an organization and how to approach the complex task of bringing about and sustaining social change.

It was at Kanthari that Tata learned of Sharma and his afforestation approach using the Miyawaki Method. Given her interest in forest conservation, Tata's mentor at the institute insisted she watch a couple of Sharma's video presentations and talk to him about her project goals.

When Tata returned home to Cameroon at the end of 2018, Buea's water crisis was noticeably worse than when she had left. In eight short months, the public taps had gone from supplying water through the rationing system a couple of days a week to simply running dry. People had no choice but to buy water from those with wells or fetch it from the forest. "It was really serious," she said, "and I wondered what we could do." She called Sharma to explain the situation and ask his advice, and he indicated that restoring dense native forest could recharge groundwater by a factor of thirty compared with the existing unforested land. "My goodness, this could be what we've been waiting for," she thought.

But there was still the question of financing the reforestation project she had in mind. Sharma referred Tata to Elise Van Middelem of SUGi, an organization that supports Miyawaki Method forest projects, and the gears began to turn. Tata first spoke with Van Middelem in August 2019, and by November her team was planting.

Out of several catchment sites needing attention, they chose to start with the Bonduma village spring on the basis that a large number of people relied on it, including the University of Buea campus and three surrounding neighborhoods. The land here was once a shallow lake throughout the dry season, but it had recently become a grassy field with a hot tub–sized pool of water at the source. Furthermore, there was space to plant—no houses or fields crowded the water as was the case at some of the other catchments.

A couple dozen people planted 600 trees comprising nine native varieties on a tennis court–sized patch of sloping land above the Bonduma spring. A few hot, dry months later, in March of 2020, the Ecological Balance team planted its second forest at the Bulu catchment in a nearby village. The difference this time around was that the rainy season was just beginning, meaning these new plants would get plenty of water as they established their roots.

The second forest is doing much better than the first one, Tata explained, because of its well-watered start. After a year of growth, the Bulu forest was 5 m (16 ft.) high, compared to 3 m (10 ft.) at the Bonduma site. "Seasons really do matter," she concluded.

Another factor in the second forest's impressive growth stems from what Tata's team learned from the first experience. "We really had challenges with the first forest . . . there were children who would play around trees and uproot them, and throw soapy water on them. One area had to be replanted over five times! It was our first trial," she explained, "and because of that we did much more education in the second community." Because of this education, children understood that the young plants were important and needed to be treated gently while they grew.

Even in the middle of the dry season, when the temperature can reach 39°C (102°F), the Bulu forest patch appeared lush and healthy, Tata said. "People stand in its shade to eat before going back into the sun." It has also shown signs of life beyond the verdant trees. "We've seen a lot of birds that we did not see before. We've also seen crabs, which are an indication of clean water. I remember the first time I saw a crab . . . I was getting emotional."

People, too, are coalescing around the beautified spring. The planting event was like a public holiday—even people in their eighties helped plant trees. The village chief attended in his traditional attire, and spoke of the ancestors' presence. "The chief and other notables are taking it so seriously," Tata said. "They say to people, 'You cannot go there and destroy it.' They are always there, always inspecting it, asking me if I am happy with the growth and whether there is anything they can do to help."

At both Bonduma and Bulu, locals have commented on changes they have observed since the native trees were planted. "We've had testimonials from the neighborhood," Tata said. She recounted what people have told her: "'Madame, we don't usually have this much water in the dry season.'

"'At the beginning, we thought you were joking [that planting trees would improve the water quality], but we've seen that the water is cooler.'

"'This thing you're doing is something serious.'"

Neighbors have also been happy to spot medicinal species they value, and have begun harvesting the leaves. "We have to beg the people to let the trees grow before harvesting them," Tata said.

The word is getting around about these reforestation projects that seem to revive water catchments, and neighborhoods are requesting them. It is hard to say no, so the Ecological Balance team always obliges, even if only to plant a few trees. For the Bonduma and Bulu plantings, tree seedlings were purchased from a local nursery. "That's one of the big challenges," Tata said, because there are not many local sapling suppliers, and none with a large selection of native plants. In addition, purchasing the saplings is half the cost of reforestation, meaning Tata has to wait for funding to become available before any new planting project can begin.

The solution? Ecological Balance started its own nursery when they realized they could not continue to rely on purchased saplings if they wanted to be able to respond to any request. The brand-new nursery is starting with 50 native species, but Tata would like to grow that number to 1,000 eventually. "The forests that we inherited from our ancestors are very rich," she said with affection in her voice.

So rich, in fact, that the Congo Basin is recognized as a UNESCO World Heritage site, owing to its incredible biodiversity. The rainforest is home to some 10,000 plant species, a third of which are only found there, and 1,000 bird, 900 butterfly, and 400 mammal species.[9] It is also home to 30 million people who depend on its provisions.[10] Unlike in Europe and the United States, where most primary forest was cleared centuries ago and replaced with a sparse and motley mix of native and exotic plants favored for their commercial or decorative value, indigenous species are easy to identify for people in Tata's community.

She said that local people can distinguish native plants "by virtue of the fact that we are forest people" who have been gathering from the forest for as far back as anyone can remember. "Most exotic trees are used as ornamentals, planted around houses and parks," she added, citing the example of eucalyptus, imported from Australia. This fast-growing plant has many uses in addition to being pretty, and has therefore been adopted widely throughout the world. Ironically, this particularly thirsty tree was planted along some of Buea's water catchments, where it actually lowered the water level.

"We've had a lot of discussion about the negative effects of eucalyptus," Tata said, explaining that there wasn't enough education about it when the plant was first introduced. Such discussions are woven into a broader series of conversations that Ecological Balance facilitates with local people on the importance of ecosystems and the link between deforestation and the water crisis.

The organization's first priority is "sensitization on the 'real' value of forest," referring to its ecological and social value. As such, Tata has put a lot of energy into talking to people, especially youth. Tata's team connects with people through the church and the university, two key pillars of local public life, visiting with youth clubs and participating in community volunteer days.

A core message she imparts to young people about solving hard problems like the dwindling local water supply is that they should not wait for the government or anyone else. "It's our responsibility, everyone's responsibility. If we don't do this now, this could become catastrophic. The future is yours, and if you don't stand up, the older people will be gone and you will be left with this water crisis."

It's a message that seems to have resonated, as Tata reported: "Suddenly youth are coming up to us and saying, 'Please, when are we planting the next forest?' It's awesome!" Tata's nine-person staff is mostly under the age of thirty; they all started out as volunteers. Hundreds of other local people have pitched in to help with tree planting and other events, suggesting that Tata's vision for the future and how to get there is catching on.

Back in 2018, she sketched out that vision for an international audience as part of a five-minute presentation for kanthari TALKS, an event that is much like a TED Talk. "Looking forward five years or more from today, this is what you will see. Can you please close your eyes for a minute?" she asked the crowd.[11]

Her talk had begun with an overview of the challenges Cameroon faces as West and Central Africa's largest timber exporter. The country is losing 30 ha (74 ac.) of rainforest every hour, she said, due to an insatiable global demand for tropical wood. The northwest part of Cameroon, where Tata grew up, has lost 63 percent of its forest, leading to both dried-up streams and flooding, depending on the season. Explaining how forest loss has also led to a decline in women's incomes all over the country, given their traditional work gathering forest products, she then outlined Ecological Balance's overlapping social and environmental programs that aim to push back the tide.

"Imagine a large stretch of land—30,000 hectares [74,132 acres] of tropical forests. See the irokos, the mahoganies, the bush mangoes thriving together. Shhhh! Listen to the birds," she called out ardently, closing her eyes, too, while everyone listened to a soundtrack of singing tropical birds and other lovely forest sounds. "Streams of sparkling water. Listen to the happy chimpanzee. It is a new day and the children are running to school. The women are rushing to the market and the men going to their farms. Ladies, gentlemen, and others, welcome to my community."

CHAPTER FOUR

Urban Oases

Concern over improving the quality of life for Parisians during the heatwaves that are unfortunately set to multiply in the coming decades is one of the key motors of our public space greening agenda.

—CHRISTOPHE NAJDOVSKI[1]

The hilltop view from the Sacré Coeur Basilica is iconic. From there, Paris is a wide, ivory-gray mosaic where you might spot a juggler on a unicycle in the foreground if you're lucky. The street view is one of dazzling relics: a Roman arena, medieval houses, a gilded opera house famed for its phantom, the burned and partially renovated Gothic cathedral home to a fabled hunchback, labyrinths of street cafés and kebab stands, thirty-seven bridges over the curving Seine River, and a few trees older than the United States. At the ripe old age of 8,000 years, Paris is not only the seat of the world's seventh largest economy, but a living archaeological site.

When IT specialist Enrico Fusto moved to Paris in 2015 from his native Italy, only one thing bothered him about the elegant city he now called home. "What surprised me," he told me in an email, "was the rarity of trees both along the grand boulevards and smaller streets, and the lack of conservation areas within the city where nature is left to develop on its own." He missed the mere presence of nature and its soothing effects amid the hubbub of human activity.

Others agree, including municipal arborist Arthur Massart of Paris's Department of Green Space and Environment, whose love for nature and lifelong dedication to trees was the subject of a 2008 article he forwarded to me, along with other documents on the importance of urban vegetation.[2] "Paris is one of Europe's densest cities, and it's very paved," Massart explained to me by phone. "What's needed are more areas that create a natural cooling effect and boost resilience to increasingly regular summer heat waves. If we cannot create a habitable environment in this city, then we have a big problem."

Culturally and materially rich, Paris attracts people: In addition to its 2.2 million residents, millions of tourists visit the city every year. Yet the City of Lights is not immune to the discomforts and dangers of the breakdown of the global climate system, and it is already suffering increasingly frequent and intense heat waves. In August of 2003, the temperature in Paris hovered between 30°C (86°F) and 40°C (104°F) for more than two weeks straight, killing 1,100 people, and in July 2019, the city hit a new heat record of 42.6°C (108.7°F).[3]

"In Paris, historically a very paved city, the heat can be suffocating during a heat wave in the areas without greenspace, which are therefore poorly equipped to defend against the heat island effect," remarked Christophe Najdovski, Paris's director of public green space, biodiversity, and animal welfare. Najdovski responded by email to several questions I asked him about the city's tree-planting, biodiversity, and climate change adaptation initiatives.[4]

Over the course of the twenty-first century, Paris's average summer temperature is expected to rise as much as 5.3°C (9.5°F) (or as little as 1°C [1.8°F]) and the number of days per year with temperatures higher than 30°C (86°F) could increase to forty-five days from the current average of ten days.[5] Rising temperatures will come with more frequent and extreme storms, flooding, and drought. A 30 percent reduction in the flow of the Seine River is expected by 2080; along with the Marne River southeast of Paris, the Seine provides nearly half of the city's drinking water.[6] Ironically, the river also poses increasing flood risks as climate change increases the likelihood of extreme precipitation events.

Paris is not alone. In 2020, for example, Montreal hit an all-time high for the month of May at 37°C (98°F), and Los Angeles County broke its all-time high temperature record, reaching 49°C (121°F) in September. On top of the 1.1°C (1.98°F) rise in the average global temperature since the 1800s, which has intensified heat waves, cities are further burdened by the urban heat island effect that Najdovski mentioned.[7] This effect is generated by the concentration of impervious surfaces, as well as heat-generating machines such as air conditioners and cars, and dearth of vegetation. When sunlight strikes concrete, asphalt, metal, and stone, it becomes "sensible heat," the heat you can feel when thermal energy is transferred to a material. On the hottest summer days, asphalt can easily burn skin.

In contrast to pavement, living plants are never hot. That's because a leaf converts sunlight into "latent heat," which we do not experience as heat. Latent heat is the energy absorbed by liquid water that turns it into vapor. Through the process of transpiration, water taken in by roots and transported up a plant's stem or trunk makes it to the stomatal openings on the leaves, vaporizing just below the surface and cooling the plant in a release of latent heat. Humans do something similar when we sweat—water evaporates from our skin and cools us.

Plants use water mainly for this temperature regulation function, while less than 5 percent of their water use is for cell production and growth.[8] An average mature tree can transpire hundreds of liters of water per day. The cooling power of every 100 liters of water transpired equates to the daily output of two central air-conditioning units for an average home.[9] Imagine what a forest can do.

Case in point: A forested park at Paris's perimeter recorded temperatures 10°C (18°F) lower than the city center on the same day.[10] Similarly, Washington, DC's wooded Rock Creek Park was

9°C (17°F) degrees cooler on a hot August day in 2018 than a treeless part of the city a few short miles away.[11] A 2006 study of the heat island effect in New York City revealed that the city's daily minimum temperature, which occurs at night, is on average 4°C (7.2°F) warmer than surrounding areas.[12]

Trees bolster a city against heat waves. And with less than 10 percent tree coverage on its city streets, according to a 2016 study, Paris had some catching up to do.[13] Street tree coverage is more than 20 percent in Frankfurt, Amsterdam, and Geneva, for example, and nearly 29 percent in Oslo.

Having pondered Paris's many shades of white, which seemed to be at the expense of green, Fusto was all ears when he discovered Sharma's online talk. The Miyawaki Method appealed directly to his nagging question about how to invite nature into the dense city. "What I liked was the accessibility of the method," Fusto recalled. "It seemed simple enough to be implemented by regular people, and effective for restoring biodiversity in urban lots of any size—even very small lots."

Fusto plunged into research on the method and even contacted Sharma and Miyawaki themselves. He also experimented with native plants in his own garden and started talking to the people around him about the methodology. Fusto found kindred spirits, and together they organized themselves into an informal collective to develop the idea of planting a Miyawaki Method forest in Paris.

By the time a call for proposals had been issued for Paris's Participatory Budget, which allocates 5 percent of city funds to resident-led projects, the fledgling group had a plan. In 2016, Fusto submitted a project proposal titled "A Sustainable, Open-Source Forest," and fellow Parisians voted for it, along with several other resident-led environmental, cultural, and social projects.

Not only were residents ready for a project like this, city leaders were, too. Paris had established its first Climate Plan in 2007 and Biodiversity Plan in 2011, both of which were updated in 2018. The 2018 Paris Biodiversity Plan outlined the context for its adoption: "Biodiversity is not just a list of species or genera, but rather, the living, interacting tissue of Earth, whose continuity, functioning and structure are as important as its composition." The text goes on to assert that emissions reductions cannot be effective for mitigating climate change without a simultaneous global commitment to protect forests, wetlands, coral reefs, and every other ecosystem, thus keeping the carbon stocked therein from being released as CO_2.[14]

In a bid to protect wildlife and shelter people from extreme heat, the city proposed to plant 170,000 additional trees, expand wetland area, remove 100 ha (247 ac.) of concrete and asphalt (about 1 percent of city surface), and open 30 additional ha (74 ac.) of vegetation to the public. A guiding objective was that nobody should ever be more than a seven-minute walk to the nearest "cooling island," which is a location open to the public such as a park, woods, cemetery, church, library, or swimming pool that is cooler than surrounding areas.

The city mapped out tree-deprived neighborhoods and streets for priority planting. And alongside the maps of cooling routes and islands for people were maps of wildlife corridors and reservoirs (both aquatic and terrestrial) within the city.

The vision was to create a web of ecologically functional habitat across the urban matrix that supported the well-being of all citizens. And, importantly, this included "Biodiversity," which was granted honorary citizenship by the city of Paris in 2016. Najdovski explained that this was a symbolic act: "This recognition contributes to consciousness-raising among as many people as possible about the risks associated with the silent disappearance of our immediate and greater ecosystems."

Once Fusto's project was approved, a site was chosen on the border of the *périphérique*, a four-lane beltway encircling Paris. Najdovski saw this ring of mostly grassy land as having strong potential to contribute to the development of wildlife corridors throughout the greater metropolitan area. A wildlife corridor is a long strip of vegetated land that connects otherwise isolated natural areas and facilitates wildlife movement and genetic dispersal and mixing. These corridors are essential for ecosystem restoration at local and regional scales.

"The beltway carves out a thirty-five kilometer [twenty-two mile] long stretch of vegetation," he said, "[and] is connected to other planted areas and reservoirs of biodiversity: woods, parks, gardens, squares, sports fields, cemeteries, the Seine River, and channels."

Najdovski called my attention to Seine-Saint-Denis, a suburb opposite the beltway in northeast Paris, which approved a plan in 2020 to increase its urban canopy from 16 percent to 20 percent by 2030. The plan emphasized tree species diversity, native varieties, conservation of mature trees, and the creation of small urban forests. To the extent that the beltway embankment provides suitable habitat for birds and insects, it is well-situated to facilitate ecosystem connectivity among adjoining pockets of natural vegetation.

With the site chosen, Fusto's team, which formalized itself into a nonprofit called Boomforest, could focus on figuring out what species to plant. Many of Paris's common or beloved trees hail from elsewhere: Lebanese cedar, Japanese pagoda, black locust, giant sequoia, and horse chestnut, for instance. The common plane tree is a hybrid rather than a wild species. Thus, such species do not make up the city's potential natural vegetation.

The City of Paris defines native plants as those that were in the region prior to AD 1500. According to the 2013 Paris Basin Native Plant Guide, "This date is characterized as the beginning of a period of great intercontinental flow of goods, notably with the first species arriving from the Americas."[15] For example, black locust was introduced to Europe from North America in the early seventeenth century, and its cultivation was encouraged for many years due to its fast growth and utility for firewood and building. However, its "adaptability and capacity of transforming ecosystem processes are the reason for its adverse effects on biodiversity."[16]

The 2011 Paris Biodiversity Plan had called for a study to identify and a strategy to start growing native plants in city nurseries. The resulting Paris Basin Native Plant Guide lists 145 plants, organized according to their size and their preferred environment, such as wetland, sandy soil, crop fields, abandoned lots, or forest.

Fusto's team tore through this document, which was rich with the information they needed; yet it was still not precise enough in terms of locality: The Paris Basin reaches west to the English Channel and north to Belgium.

To whittle down their list to only the most locally adapted species, the Boomforest group cross-checked the Paris Basin plant list with online guides describing the ecological conditions to which each tree or shrub is suited. They also consulted local botanists, the city nursery director, and Miyawaki's team. This iterative process eventually resulted in the selection of thirty-one native trees and shrubs, including a few species each of maple, oak, and linden, as well as hazelnut, common beech, holly, elder, and wild rose, apple, and cherry, among others. "The list is made up of species that live well together, thus favoring the resilience of the vegetation stand as a whole," Fusto explained.

On a cool March morning in 2018, forty volunteers planted 1,200 saplings into 400 m^2 (4,306 ft.^{2}) of soil beside one of Europe's busiest freeways, at the Porte de Montreuil. A year later, they planted another mini-forest a mile north of the first one, also on the beltway. The Paris Tree Plan released in October 2021 articulates a goal to "transform the beltway into a circular forest" to the extent possible, while also experimenting with urban forest planting approaches elsewhere in the city, using the Miyawaki Method or otherwise.[17] In the meantime, Boomforest has planted mini-forests at a Paris community garden, in a clearing within a degraded suburban woods, and in the city of Lyon, all the while consulting groups throughout the country on the Miyawaki Method.

Fusto and I met on a rainy afternoon in late 2020 to visit the first beltway mini-forest. He unlocked a gate behind a sign demarking the site as a Participatory Budget investment, and we walked up to the thicket of tall and mostly leafless deciduous trees growing on a meter-high oblong mound. The floor of the dormant forest was green in the mild Paris winter on account of grass and other small herbaceous plants growing among the trees. A few evergreen shrubs lent shades of deeper green.

Walking around the stand of young trees, we chatted over the roar of traffic. "This coming year in March of 2021, the forest will be three years old," Fusto announced, as if speaking of a child about to start preschool. "So, we're entering into the famous period of autonomy. We'll soon see if it's true that we no longer need to maintain it."

During the first three summers, small teams of neighborhood volunteers weeded the incipient ecosystem on a monthly basis; the city watered the trees only once or twice each year. Not all weeds needed pulling out, though. "The bindweed, thistle, and black locust needed to be weeded, but not the dandelion, which is too short to bother the young trees and shrubs and is quickly shaded out anyway." Actually, he said, pointing toward the middle of the grove, "you can see that even the bindweed is no longer growing in the middle, but only at the more exposed edges." Over the previous summer, Fusto added, this older plot needed less weeding than the younger mini-forest that had been planted a year later.

A further sign of this little ecosystem's growing maturity is that insects and other small animals are showing up. Fusto has seen bees, European firebugs and ladybugs, mole burrows, and a pigeon nest. There are also lots of earthworms under the loose and spongy forest floor, especially as compared to the adjacent grassy stretch, where Fusto found little visible evidence of life when he took soil samples.

When Fusto and I spoke again in August 2021, he said the last weeding had been in March—pretty much at the mini-forest's three-year anniversary. The trees are in very good health, he said, and "most are taller than us." Furthermore, the vegetation layers—ranging from shrubs to understory and overstory trees—were now noticeable, even though some won't reach their full height for another fifteen to twenty years or more. When they were first planted, the plants were uniformly less than a meter tall. Even one year prior, Fusto said, the layers were not yet quite apparent.

While weeds were no longer present in the increasingly shaded area, a few forest species (beech, oak, and gooseberry) were sprouting new plants around the edge of the forest. It is not that last year's two-year-old trees had already produced acorns or beechnuts, but rather that these seeds had arrived from nearby trees outside of the mini-forest and the seedlings had been allowed to take hold—not mowed, as they would have been elsewhere. The gooseberries, which take fewer years to mature, may have self-seeded from the mini-forest.

By the winter of 2021, Boomforest hadn't yet added another beltway mini-forest to the original two, although plans for a third were in the works. The association was busy responding to other requests. In November 2021, they applied the Miyawaki Method in front of a twenty-eight-story suburban apartment building in Colombes, a small city tucked into a bend in the Seine River northwest of Paris. The mayor, who planted a few trees himself, explained to the media that Colombes would likely plant several more mini-forests as a way to help combat climate change.

Boomforest's initiative is gaining traction one local hero at a time, resulting in the sporadic placement of tiny ecosystems rather than the strategic development of an ecosystem corridor—at least for now. As the hyper-local landscape transformations prove themselves over time, though, perhaps the Miyawaki Method will become a centerpiece of Paris's ostensibly biodiversity-sensitive landscaping strategy.

An Earthy Education

Around the time Fusto was developing his mini-forest proposal in Paris, Daan Bleichrodt was leading a few classrooms of schoolchildren in planting a tiny native forest beside a busy road in Zaandam, Netherlands.

Bleichrodt works for IVN Nature Education, a Dutch organization with a network of 25,000 volunteers that facilitates nature experiences and education. This planting day was the culmination of a dream Bleichrodt had been brewing for a year or two. It started while he was leading a program that organized daylong expeditions for urban youth to visit forests and beaches. He had been stunned to discover how many children who live only a few miles from the coast had never seen the ocean.

"So many kids grow up without nature," he said to me in a video call. "That's scary to me because we have all these big challenges—like climate change, loss of biodiversity, and the plastic soup. Will you protect nature as an adult if you didn't grow up with it? In 2050, today's kids will be at least thirty years old, and we need to be carbon neutral by then. So there's a great urgency to connect kids to nature."

The Netherlands is one of Europe's more urbanized countries, with nearly three-quarters of the population living in a greater metro area.[18] When built infrastructure is all one sees, it is easy to feel separate from the web of life. The day trips were successful in providing fun, enriching experiences for city kids, but it bothered Bleichrodt that many would likely not return simply because their families lacked the means. This got him thinking about other possibilities.

He had heard about Cruyff Courts, a concept created by Dutch soccer star Johan Cruyff. The courts are small, fenced-in soccer fields with artificial turf, installed in neighborhoods with no other safe outdoor places to play. The idea is to encourage kids to play outside without the worries of traffic-clogged streets. Many dozens of courts have been built throughout the Netherlands and elsewhere, visited by some 65,000 kids weekly, according to the Cruyff Foundation website.[19]

Bleichrodt wondered: What if we could bring nature to city kids in the way Cruyff Courts bring them soccer and outdoor play spaces? Then he stumbled upon Shubhendu Sharma's talk about the Miyawaki Method. It occurred to Bleichrodt that this was a way to create schoolyard-sized ecosystems that kids, parents, and teachers could observe on a daily basis as their mini-forest grew and changed over the seasons and years.

Fascinated, Bleichrodt described the amazing methodology to friends and colleagues: how by densely planting multiple native species in well-composted and mulched soil, the trees grow vigorously and the system quickly becomes richly biodiverse and self-sustaining. Perhaps skeptical of the then unknown technique, his audiences did not share his enthusiasm at first.

He realized that he needed to take a step back, and rather than present the information as fact, frame the mini-forest idea as a series of questions so that people would be more receptive. Could this compelling methodology that works in India and Japan also possibly work in northern Europe? Could trees planted this way grow ten times faster than those planted by conventional methods, as advocates claimed, and foster 100 times the biodiversity? There was only one way to find out. Bleichrodt contacted Sharma to learn more about the method. He also started looking for a city in which to plant IVN's first mini-forest.

His first thought was to propose a tiny forest in a city called Haarlem, just west of Amsterdam, on the basis of its reputation for being the most paved-over downtown in Holland. A boost of vegetation in such a place would not only be pretty but also absorb floodwater and cool and clean the air in a high-traffic zone.

Hoping to convince Haarlem's municipal staff of such benefits, Bleichrodt invited them on a bike tour to look at potential forest sites. City staff canceled at the last minute, so Bleichrodt and a friend visited the eight locations on their own. Their favorites included a spot near a government building beside a busy road, a playground in a low-income neighborhood, and a small empty lot near a shopping center. Bleichrodt reported his site recommendations to city officials, who responded within minutes that none of these spots were possible. Case closed.

Despite Haarlem's ample paved surfaces, the ancient, culturally rich "flower" city (so called for being the historical center of the country's tulip-growing region) has no problem attracting residents.

It is "beautifully located," Bleichrodt said, as it is near the beach, a forest, and the capital city of Amsterdam. Maybe a less popular city than Haarlem would be more receptive to new ideas, and motivated to make the city more attractive for its citizens, Bleichrodt's co-worker suggested.

Located on the outskirts of Amsterdam, on the Zaan River, Zaandam is a small, industrial city and historical lumber port with a track record of trying new things: In 1971, Zaandam broke ground on Europe's first McDonald's restaurant.[20] A tiny forest proposal here turned out to be right on target: Bleichrodt first approached city leaders in May 2015, and by December he was planting one of the continent's first Miyawaki mini-forests.[21]

The journey from May to December was not without its bumps, though. Once the municipality of Zaandam was on board with the basic concept, there was still a steep learning curve ahead for everybody involved. The Miyawaki Method presented a completely new approach to planting trees. Why plant so densely? What combination of native species should be planted, and in what proportions, to re-create the layers of a natural forest?

People were enthusiastic about the project's potential, but they needed to understand it better. Forest planting was still new to Bleichrodt, and he felt unprepared to answer to the skepticism coming from every direction, even from other IVN staff and volunteers. The answer to a question like: "Why must we add 5 to 10 kg [11 to 22 lbs.] of compost per square meter?" had to be more robust than "because it's good for plants."

Bleichrodt invited Sharma to Holland to lead the charge. The mere presence of an international expert who had already successfully planted dozens of mini-forests using the Miyawaki Method gave credibility to the new project. However, as Bleichrodt recalls, some skeptics saw it as a waste of time and resources to invite an expert from abroad to advise locals on afforestation when the Netherlands had its own experts.

Sharma arrived a month before the planting date. He spent a week in Zaandam, where he presented his afforestation concept at city hall to an audience of 150 people, including municipal authorities and local residents. He met with landscapers, visited the site, took soil samples, and visited nurseries, all the while fielding questions and explaining the details and rationale for the novel technique.

Some skittishness remained even up to the planting day. In fact, the project was almost derailed when buried debris was unearthed at the planting site. But a brave civil servant defended the project, Bleichrodt said, and in the end the site was simply moved over to a new patch on the same strip of lawn.

Five years later, the Zaandam tiny forest is a dense emerald explosion more than 12 m (39 ft.) tall, and home to 595 plant and animal species and counting.[22] Wageningen University researchers expect this number to continue to rise for a few years, after which it might decline as the canopy fills in, limiting the sunlight that reaches the forest floor. Some flowering plants will disappear, along with their accompanying pollinators, but different species and bigger animals will likely show up.

Local residents volunteering as "citizen scientists"—nonprofessional scientists who undertake scientific research projects (in this case, wildlife monitoring)—are partly to thank for these findings. Starting in 2017, a small team has made monthly visits to the young ecosystem to count species and report the numbers back to Wageningen researchers, who themselves were gathering data too. The team has found increasing numbers of species every year.

Even with the Zaandam tiny forest in the ground and growing well, IVN's fledgling program still struggled for recognition in its first few years. The group didn't plant any tiny forests in 2016, and only five took root in 2017. However, like young trees expanding their root systems in winter while, on the surface, it appears that nothing is happening, Bleichrodt was busy doing the fundraising groundwork to build and sustain the new program.

Then the buds of the Tiny Forest program began to open. In 2018, IVN received a whopping €1.85 million ($2.07 million) donation from the Dutch Postcode Lottery, a national lottery that donates half of its proceeds to charities. This grant meant more than simply its infusion of resources; it was a giant stamp of public approval. The same year, Wageningen published its first report showing the strong biodiversity effects of the Miyawaki Method in Zaandam.

Henceforth, cities across the Netherlands started requesting tiny forests. IVN decided to set up an application process to narrow the field of municipalities. Out of sixty-six applications in 2018, the nonprofit could accept twelve. The following year, they again received sixty applications and approved thirty, with each city committing to plant at least three forests.

By the spring of 2021, Bleichrodt and his now forty-person team at IVN had planted 145 tiny forests in collaboration with cities, in addition to another 60 on private property, including one at a nursing home in Haarlem. Each tiny forest the size of a tennis court costs a city €22,000 (about $25,000), half of which is covered either by the Dutch Postcode Lottery funding or through provincial government support. This covers materials and labor, including professional soil preparation and IVN's intensive engagement with communities and schools before, during, and after planting day.

While this is a lot of money per mini-forest, its significance shrinks in the context of a municipal budget. Even for a very small city with a budget of €20 million (more than $22 million), the total cost of a tiny forest would amount to a one-thousandth of the annual budget. Many cities have much higher budgets, ranging in the hundreds of millions or billions. For big cities, park and recreation spending alone pushes into the hundreds of millions.

Once a city is chosen, the next step is selecting sites. IVN asks local residents and schools to apply for a tiny forest for their neighborhood. Applicants then present their vision and project idea in a public meeting. "The first meeting we did had ten people pitching their motivation for a forest. You get these beautiful personal stories," Bleichrodt said.

After the sites are selected, a second community meeting is held to discuss tiny-forest design and to identify a local school or daycare center that will adopt it. Sometimes the school is the initiator, in which case community partners are identified to help with the project. Either way, every city-sponsored tiny forest belongs to a school or a daycare center. The next step is to decide which group of kids will do the planting, and to plan a preplanting guest lesson in their classroom.

Arriving in character as forest rangers, IVN instructors make a plea to the kids: "Can you please help us create a forest? It's no fun to be a ranger without a forest." This introduction is followed by a lesson on forests in anticipation of the upcoming planting day.

Meanwhile, teachers are trained in how to lead activities and lessons outdoors and are provided with curricula for students ages four to twelve years. IVN focuses on this age group, Bleichrodt explained, "because we know from experience and from research if you manage to forge a connection with nature before the age of twelve years, you will be connected for life."

"The planting day is the best day," Bleichrodt said. "There's a lot of excitement, enthusiasm, and pride." He described how kids really seem to feel proud about their forests and about the particular trees that they planted and continue to take special care of. "That sense of ownership—that's what we've tried to create." That and a sense of wonder, he added, which is what happens whenever a child finds something as delightful as a spider while exploring her forest.

A few months after planting, students measure the trees and look for insects and other creatures in their forest. One age group in the school is designated the "rangers" and is responsible for forest maintenance, including picking up litter and leading forest tours for friends and family. Each forest includes an adjacent space equipped with log benches for social activities, lessons, and even barbecues and birthday parties.

Bleichrodt says that the way each community interacts with a tiny forest varies. While the original Zaandam forest never became a community meeting site, it is perhaps the most biodiverse by virtue of its being left alone. Its few and infrequent visitors (the citizen scientists) are focused entirely on the forest's well-being and tread very lightly in it. By contrast, the tiny forests planted on school grounds are subject to compaction, regular twig-breaking, and fort-building. Kids connect with the forest by going into it, touching it, and looking for bugs. Both playtime and outdoor classroom lessons provide this kind of experience, but these activities also result in some damage to the young and still-fragile forests.

It is a small trade-off to make in exchange for what is gained, Bleichrodt thinks, when children enjoy their young tiny forest. "That's what I wanted to do: to connect kids growing up in an urban environment who don't ever go to nature—to get them to plant a tree, watch it grow, be a ranger of their own forest, develop a love for nature, and help to become the nature restorers of the future. Those are the places I like the most."

Bleichrodt added that it's the well-trodden forests that may endure the longest anyway. The Zaandam forest was almost bulldozed once by city staff concerned about underground pipes nearby. It was one of the citizen scientists happening upon the scene who stopped it before any trees were uprooted. "If the forest has no meaning to the community, it's not going to have the proper protection," Bleichrodt reflected, "as you can see in Zaandam. But fortunately, our volunteer Fred saved the day."

Tiny forests are selling like stroopwafels in the Netherlands. The country is well on its way to IVN's goal of 230 tiny forests by the end of 2022. Even architecture firms are contacting IVN to request tiny forests. "Over the past three years," Bleichrodt said, "I've been getting calls from architects and housing project developers who want to pay attention to biodiversity." He said that some developers are designing their outdoor green infrastructure even before planning the houses, reflecting a shift in consciousness and priorities.

Bleichrodt attributes the program's success partly to a new policy requiring Dutch cities to take "climate stress tests" every six years and then to make improvements, such as planting a few tiny forests, to increase local resilience to extreme weather.

Another factor in the Tiny Forest program's popularity is growing public awareness of the climate crisis. Several recent droughts are raising alarm in a country long dedicated to chasing away water.

The global "insect apocalypse" also captured news headlines in the Netherlands with the 2017 release of results from a twenty-seven-year study conducted in Germany by a network of citizen scientists.[23] This group observed a 76 percent decline in flying insect biomass since 1989—a dramatic reduction in the sheer volume of insects. A 2019 study estimates that 40 percent of the world's insect species will be gone within a few decades due to habitat loss, pesticides, pollution, pathogens, and global warming.[24]

The loss of insect populations is creating a ripple effect up the food chain. Bird populations in the French countryside have dropped by a third in less than twenty years, with some species' populations falling by 70 percent. Researchers say this loss is largely due to pesticide use that devastates the populations of insects on which these birds feed.[25] Across Europe, 13 percent of bird species are threatened with extinction.[26] Since the 1970s, the bird population of North America has declined by 30 percent due to habitat loss and other factors.[27]

While awareness of these crises heightens people's sensitivity to environmental issues, the situation can seem too overwhelming to trigger action. "The crisis is one of empowerment," Bleichrodt reasoned. "Because of global capitalism and multinationals, people feel powerless. It's not a crisis of the unwilling, because three-quarters of people have a 'hippie' view of nature. Environmental organizations need to look for new ways to empower people if we are going to tackle this crisis."

In other words, most of us appreciate nature and would like to protect it, but we get drawn into an economic system that is constantly urging us to use nature as if it were nothing more than an endless supply of disposable resources. Even our moments of clarity about the true costs of the economic system we participate in are blotted out by a sense of its intractability and our powerlessness in the face of it.

Planting tiny forests is one way to restore a sense of agency in individuals and communities. Now regularly flooded with requests from people within the Netherlands and around the world wanting a tiny forest, Bleichrodt and his IVN team are slowly shifting toward a "train-the-trainer" model for the Tiny Forest program. "For the next twenty or thirty years there's going to be an extreme demand for planting forests. We just have to make sure that we get a larger community of people who can work with us. That will require us to be training and helping others to do it." In an encouraging example of what's possible, an IVN volunteer who planted a tiny forest in her backyard in 2017 went on to train 100 others to do the same.

"I think we've just started with reforestation," Bleichrodt affirmed, citing Dutch plans to increase forested area by some 37,000 ha (91,400 ac.) by 2030.[28] He stressed, however, that such goals must be approached not only in terms of hectares and numbers of trees, but in terms of the creation of self-renewing ecosystems.

CHAPTER FIVE

Campuses, Coastlines, and Foothills

The process of restoring green environments to residential and industrial neighborhoods, as well as to spaces in and around transportation and other public facilities, cannot be limited to mere decoration.

—Akira Miyawaki and Elgene O. Box[1]

"White oak, white oak, white oak!" call out hundreds of manufacturing workers, managers, and their families in Madison, Indiana, as one holds up an exemplar seedling like a trophy. This is part of a "naming ceremony," the signature of Miyawaki's tree-planting festivals wherever they are held, from Japan to Kenya to China to the United States.

Over the course of six years, Arvin Sango employees and local community members planted a total of 22,000 native trees and shrubs on long strips outside the automotive parts plant in Madison. The company's annual event and the intensive preparation leading up to it is documented in hundreds of pictures and videos on the event's social media page. If visitors scroll far enough down the page, they will find shots of Miyawaki at the first planting in 2014.

"We are here today because we love the environment, Arvin Sango, and our families, and we want to keep every day like today: sunny, breathable, and pleasant. Trees help do that," says a participant flanked by his family at a 2018 event. Another planter interviewed in an online video is motivated to take care of God's creation, while a third simply says that it is the right thing to do, adding: "Trees, what's not to like?!"[2]

The southern Indiana facility is north of the Ohio River and south of a 20,234 ha (50,000 ac.) wildlife reserve, but its more immediate surroundings are cropland, warehouses, and other industrial facilities. "There's not a whole lot of pretty things in the area, but the trees stand out," said Dan Grady, who organized the planting events, adding that the thin swaths of forest also block out noise from the road.

A training manager for Arvin Sango, Grady has a calm demeanor and an air of competence. He struck me, when we video-chatted, as someone who cares about doing a good job whatever he undertakes—whether coordinating manufacturing or eco-restoration activities. You would have to be highly organized to orchestrate the planting of 4,000 trees in less than two hours by 400 adults and children. And planting trees is only half of what makes up an Arvin Sango Tree Planting Day.

Images of face-painted kids, hamburgers on the grill, an ice cone stand, a rescued-birds-of-prey exhibit, a petting zoo, hot-item raffle prizes, and a high school band performance indicate that these events have been tailored for family fun. Convincing as many as five hundred people to show up before 9 a.m. on a Saturday to plant trees, sometimes under a light drizzle, calls for a bit of sweetening up, and Grady needs to attract enough volunteers to get the job done.

But once that task is accomplished, Planting Day is full steam ahead. As the crowd assembles in the still-early morning, kids visit bunnies, reptiles, and owls while their parents check in. The event then kicks off with the tree-naming ceremony, which features the mayor, a few company executives, and a lucky kid from the audience who introduce the names of the main species that will be planted.

"They walk up to the microphone, and they raise the tree, and they say the tree's name three times," Grady explained. Everybody in the crowd repeats the tree's name each time it is called out. "It's to introduce the people to the trees, so that they do not just memorize the name but are also able to identify it. It's to encourage a relationship with the trees themselves," he added. To aid with the identification lesson, potted specimens are lined up on tables in front of the stage with labels displaying a photograph of the tree and its name.

Next, Grady takes the stage to demonstrate the planting technique as per Miyawaki's instructions. Using a model, Grady explains how far apart and how deep to dig the holes, how to mulch around the small plants, and how to flag the smallest ones to avoid their being stepped on. Finally, he demonstrates a special knot used to secure twine around stakes. The twine will stitch down the straw mulch after planting and keep it from blowing away (see "Planting," page 144).

At last, the army of planters, loosely assembled into about twenty groups, advances on the prepared mounds. Seedlings, shovels, and straw bales are ready to grab. The pretrained group leaders demonstrate the planting procedure on site, and the battalions are released. Toddlers, teens, and grown-ups alike form a veritable hive of coordinated activity, everyone absorbed in the work.

An hour and a half after it starts, the planting is already finished. People de-mud their boots with scrubbers and tubs of water. The crowd then drifts back to the big white tents for a free lunch and the calling of raffle winners.

If the events flow without a trace of chaos, it is only because meticulous preparations filled the preceding days, weeks, and months. For Grady's part, preparation started when he traveled to Miyoshi, just outside Nagoya, Japan, in late 2013 to participate in a planting festival and observe its flow. Arvin Sango is a subsidiary of Nagoya-based Sango Co., Ltd., and it was this parent company that proposed the Miyawaki tree-planting effort for the Indiana facility.

"At the time, I really didn't know much about tree planting, so I didn't have many expectations," Grady said of his visit to Japan.

"A year or two before we planted at Arvin Sango, Miyawaki came to give a speech to the company members, and we also invited the public to attend. So, it was at that time that I got the first inkling of what the Miyawaki Method was."

In Miyoshi, Miyawaki led the leader training on the eve of the event. He demonstrated how to handle the seedlings and had participants plant a few trees on their own to get the feel of it. "He was taking us through what he expected us to do with the volunteers the next day," explained Grady, who participated with help from a translator. He recalls flying home afterward feeling slightly overwhelmed. "They had all kinds of ways not just for getting the site ready for the planting, but setting up the registration, setting up the gift bags that they would give the kids. I knew we were going to be expected to do the same and be just as organized."

Back in Indiana, Grady became an ambassador of sorts for the Miyawaki Method as he promoted Arvin Sango planting events and conducted leader trainings. When presenting the project to local gardening groups, he got some pushback. "You can't plant trees that close together, it's not going to work," people would say.

"But yes, it does work," Grady said. He has seen firsthand how well Arvin Sango's forest strips are growing. Pictures archived online show the landscape's transformation in Madison. In the 2014 images, people and planting mounds are situated in a wide, tidily mowed lawn with an asphalt road, a water tower, warehouses, and trees in the more distant background. In pictures from the final 2019 planting, by contrast, volunteers stand before a lush wall of forest that has grown up since the previous years' plantings. These older stands, which grew at a rate of three feet per year, tower overhead. The interiors of the stands are fully shaded, a luminescent green dome overhead held up by hundreds of still-thin trunks just a couple of inches in diameter.

"Sometimes it's hard to change people's mind about how trees should be planted," Grady reflected. "Typically, when they plant trees, they plant a couple of pine trees and they plant them pretty far apart. They're not planting trees to make a forest." The whole experience has made Grady himself more attuned to reforestation techniques. He suspects that if more people knew about the Miyawaki Method, more would be using it.

A Sango Story

Arvin Sango's parent company started planting Miyawaki forests in 2006. Earlier that year, the president of the Japanese automobile parts manufacturing company had heard a compelling talk on national TV: It was Miyawaki explaining his method and telling the story of his first forest, constructed with Nippon Steel Corporation in 1971. When a Nippon Steel representative first contacted Miyawaki's lab at Yokohama National University requesting a forest be planted at their new factory site, Oita Steelworks, Miyawaki was skeptical about the businessman's sincerity. Miyawaki had no interest in working with companies that were not serious about ecosystem restoration, and he was also concerned about the potentially harsh conditions for young trees at the site due to pollution from the factory.

Evoking the senior status of native trees, he told Nippon executives: "The primary trees for potential natural vegetation like these species of chinquapin [*Castanopsis*], machilus [*Machilus*], and oak [*Quercus*] have grown together with the residents of this region over hundreds of years. I want a guarantee that if the trees of these species that we planted at the steelworks all suddenly die off one day, you will turn off your blast furnaces."[3] The Nippon Steel team asked for three days to consider this deal, finally returning with a promise to control the pollution at its source as much as possible to avoid stressing the trees that would soon take root there. The answer seemed sincere to Miyawaki, and the project went ahead.

Impressed by Miyawaki's tenacity, Sango's president, Kozo Tsunekawa, promptly contacted him. At the time, Tsunekawa was facing the problem of toxic chemicals that had been detected in the ground at the company's first plant in Nagoya—Sango was charged with cleaning it up. The company turned this problem into an opportunity to make bigger changes. In the same year that Sango planted its first small forest with Miyawaki, it also explicitly incorporated social and environmental principals into its guiding philosophy.

A 2018 Sango annual report states that its goal is to "become a company that contributes to society, is trusted, and grows sustainably." The text continues: "Although this is nothing new and extraordinary, recently, we have seen a number of companies lose the trust they have built over many years in an instant, because they failed to observe this basic rule."[4]

Indeed, the daily news is full of examples of large companies breaching public trust—shirking responsibility for having polluted a local community, failing to keep workers safe, or marketing a superficial gesture like planting a couple of trees as an unwavering commitment to the public good.

Sango's words suggest that it is possible for a company to choose to be a responsible member of society—a choice that requires daily renewal. A few facts suggest Sango's efforts are genuine. For one thing, the company has little to gain sales-wise from planting trees, given that its customers are other manufacturing companies rather than the general public.

Second, the company went far beyond a single token forest at its corporate headquarters. In 2011, having planted mini-forests at most of its sites in Japan, Sango gathered the leaders of its overseas facilities to meet with Miyawaki and then take an oath to plant forests on their respective grounds. In addition to Arvin Sango in the United States, plants in Turkey, Thailand, Indonesia, and China have planted forests. Sango aims to get a total of 350,000 trees, all native to the sites where they are planted, in the ground by the end of the multiyear project.

In preparation for the inaugural planting in Nagoya in 2006, Miyawaki led company representatives on a tour of the nearby Atsuta Shrine forest to identify native species. And just as he had laid out conditions for working with Nippon Steel three decades earlier, Miyawaki insisted that all of Sango's top leaders be active participants in the planting and be willing to get their hands dirty.

Speaking to me over a video call from a Sango conference room, Sango spokesperson Kyoko Ikegami and senior advisor Yukio Goto explained that relations between managers and employees in Japanese companies tend to be quite formal, and that it is a rare CEO who would consent to dirtying his or her clothes in public. But such a willingness was precisely what Miyawaki's teaching method required on the part of corporate leaders.

"He would tell the top management, the president, boards of directors that they need to get involved with their own hands," Ikegami said. "He showed our bosses how to create little forests, and why we need to do so. By seeing those actions—not just words, but actions—I guess we as employees are convinced that tree-planting is something important."

In addition to Sango and Nippon Steel, Miyawaki worked with Mitsubishi, Toyota, and Yokohama Rubber, among other companies. There were economic reasons, too, for a business to collaborate with Miyawaki. In *The Healing Power of Forests*, Miyawaki told the story of a power company in southwest Japan, where for twenty years employees had weeded, pruned, and thinned a monoculture stand of trees ultimately intended to be sold as timber.[5] Over that time, however, global wood commodity prices fell and a typhoon damaged the stand; in the end, the trees could not be sold. This company was attracted to Miyawaki's offer to plant new trees that would cost little or nothing to maintain, would resist storm damage, and would last indefinitely.

After Nippon Steel's first Miyawaki forest at Oita Steelworks, the company planted several more. With their long taproots, the trees in these forests have stood tall and strong through earthquakes and typhoons. "Thirty years after the initial planting," wrote Miyawaki of the forest he planted in 1976 at Nagoya Steelworks, "there was a green belt of trees twenty to twenty-five meters tall, and all the buildings and access roads were surrounded by thriving environmental protection groves."[6]

Through business partnerships, Miyawaki has been able to reach hundreds of thousands of people with his ecological philosophy and method, and on his own terms. His success seems to stem from his insistence that project partners be as serious as he is about the integrity of the method. Leading by example, the botanist himself worked at the events he led well into his eighties, stooping down alongside everyone else to plant seedlings.

Forest Seawalls

The magnitude 9.0 earthquake that triggered a tsunami on the afternoon of March 11, 2011, struck barely offshore on Japan's northeastern coast. Video footage taken with a handheld camera from the upper floor of Sendai Airport's passenger terminal shows a sheet of brown water galloping toward the airport and then surrounding it. As the water rises, cars, airplanes, and broken wood planks bob and bump around in the turbid water like toys amidst eddies of unidentifiable jetsam and flotsam.

The city of Sendai, a few miles inland, was spared from flooding, but the airport is right at the coastline, where the water swelled to 8 m (26 ft.). Today, a monument marking this height stands between the rebuilt airport and the ocean at the gateway to Millennium Hope Hills. This newly created park in the small coastal town of Iwanuma stretches along about 11 km (nearly 7 mi.) of coast south of Sendai. The park is a memorial to the lives and homes lost in 2011, a bulwark against future tsunamis, and a Miyawaki Method demonstration site.

On March 12, 2021—by chance ten years and one day after the tsunami—I spoke with Yoshitomo Takano, who works for Espec Mic, the company that oversaw the forest planting at Millennium Hope Hills. Takano was still a child when he met Miyawaki during his 1993 retirement lecture at Yokohama University. Takano's father, Yoshitake, worked for the Ministry of Construction and had partnered with Miyawaki on numerous infrastructure projects. He introduced his son to the retiring professor just after the lecture.

"When we entered one of the classrooms, there was a small, old man who was wearing a white lab coat," Takano said. "He stood up and shook my hand, saying, 'Thank you so much for coming today.' I was overwhelmed by the pressure I felt from his big hand, and by his big eyes." Flustered by the professor's welcome, the young Takano forgot the greeting he had prepared.

Though still too young to take much interest in Miyawaki's philosophy that day, Takano could appreciate the forest it had manifested. Miyawaki and colleagues had planted along the main road into the Yokohama National University campus in about 1975, so by the time Takano first walked through the alley of trees, it was nearly twenty years old.

He recalled trailing behind his father as they hiked up the sloping path from the nearest train station to the campus. Struck by the sight of the evergreen foliage around him, he commented: "They took such good care of the trees when they built this university, didn't they? Because they have built the buildings and roads without disturbing the trees." Takano's father looked confused for a moment, then laughed and explained that the trees had been planted after the university campus was built. "The trees there were so big and they were standing as if they had always been there," Takano said, "so I hadn't imagined that they had been planted recently."

Beyond their size, these trees had a peculiar aspect to his eyes. "My image of forests used to be of something pricklier," Takano said. The woods he knew as a child in greater Tokyo consisted of pines, cedars, and some deciduous species. It was rare to see native broad-leaved evergreen forests in this area. He remembered a distinct coolness as he entered a building shaded by the thick grove.

Takano did not give much thought to the retired professor or his forests for many years after that day, apart from feeling a mild annoyance that his dad always seemed to be drawn away from home on projects with Miyawaki. After graduating from school, Takano went to work for a construction company, following in his dad's civil-engineering footsteps.

Over the course of thirty years, the elder Takano collaborated with Miyawaki on about the same number of projects. He was in charge of building national roads, which during the 1980s started to face opposition from local communities concerned with air pollution, noise, and groundwater depletion associated with the construction projects.

For example, the younger Takano recounted, "The construction of Kashihara bypass road in Nara Prefecture had been halted for ten years because of opposition from local residents." The integration of thick Miyawaki-style forest bands buffering neighborhoods from the new road turned out to be key to proceeding with such projects. The forest bands blocked out noise and the smell of exhaust. In some cases, when a new road was laid through an already partially degraded landscape, the promise that a hearty native forest would soon grow in an area cleared of flimsier secondary vegetation seemed to some like a reasonable compromise.

Over time, the younger Takano tired of his daily construction work, which involved constant tree cutting and vegetation clearing. For respite, he joined his dad at a few planting events organized by Espec Mic, the company that has participated in about half of Miyawaki's 2,785 projects in Japan. Finding that he quite enjoyed the landscapers' company, he became a regular volunteer planter until he was eventually invited to formally join the team. Today, he supervises soil quality and the preparation of planting mounds for Espec Mic.

Among many hundreds of Miyawaki Method groves the company has planted in Japan and abroad are four "forest seawall" projects along the Pacific shore, including the forest at Millennium Hope Hills. These are specifically designed to protect coastal areas from the occasionally savage ocean.

Takano spoke to me from the top of one of the park's fifteen hills, each created by the mounding up of debris collected from the surrounding area and topped with sediment deposited by the tsunami. Takano pointed to a five-year-old Miyawaki forest. Speaking to me through an internet link, an interpreter, and against the microphone-filtered rattle of wind coming off the ocean, his explanations were brief. At that moment, it was the view that counted.

The top of the hill was grassy, as was much of the surrounding area—the other fourteen hills and the space between them. The vista is far from spectacular—that is, unless you consider it in its historical context. Ten years prior, almost to the day, houses, farms, and everything else on this stretch of coast was abruptly swallowed by the sea.

As viewed from Google Earth, the area beyond the park contains clusters of building and house foundations surrounded by bare sandy ground and scattered weeds and scrub. Here and there, structures such as skate parks, playgrounds, cemeteries, greenhouses, and solar farms appear to have been installed post-tsunami in the space between the ocean and the more habitable inland areas.

Off to the right in the scene Takano shared with me, I could see a path tucked in between solid green strips of densely growing young trees. This forest embankment path weaves along the length of the park, connecting its hills. In the event of a future tsunami, the hills are meant to serve as high shelter points, and the forest embankments on either side of the path will slow down the water and prevent cars and other debris from being carried out into open ocean by a retreating tide.

"Right after the tsunami, I saw so much debris here," Takano said. He visits the park each season to check on the plants. He has been encouraged to see them growing steadily in spite of ever-present strong and salty wind, occasional frost, and an initial overpopulation of mice and rabbits that damaged some of the seedlings' roots.

"With construction going on all the time along the coast, I can see human recovery of infrastructure on one hand, but here at Millennium Hope Hills I can also see that nature is recovering," he said. Mice and rabbits, which were the first animals to rebound after the tsunami, have been brought into balance thanks to the return of northern goshawks, grey-faced buzzards, and eastern marsh harriers—all birds of prey. Takano has also spotted a fox.

"Red foxes are at the top of the forest ecosystem in Japan, as wolves do not exist here anymore," Takano explained. "I was so moved to see the fox just two days before the ten-year anniversary of the Great East Japan Earthquake—to witness how far the ecosystem has recovered as the result of our forest creation work."

Doryu Hioki, who spoke to me from his office on a three-way video call with Takano, was similarly happy about the fox sighting. Hioki is a Buddhist monk and president of Mori no Bochotei Kyokai, or Forest Seawall Association (FSA), the organization that created Millennium Hope Hills. "Rich forest means rich animal life and rich soils, so many animals and birds gather in a rich, native forest. It's a kind of evidence," Hioki said. He has seen foxes, birds, and insects in the quickly maturing Miyawaki forest at his own Rinnoji Temple in Sendai, as well.

Hioki discovered Miyawaki and his method in 2003, after having been informed by city officials that 507 cedar trees lining the walkway to his temple would be cut due to construction of a highway tunnel underneath the walkway. The convening of the third international climate conference in Kyoto, Japan, a few years prior had gotten Hioki thinking about climate change, prompting him to establish organic gardens and a composting system on the temple grounds. The idea of losing so many trees concerned him, so he started to read books about forests, one of which was Miyawaki's *Plants and Humans* (printed only in Japanese).

Hioki was struck by the book's discussion of the difference between "real" and "artificial" forests—a distinction he had never previously considered. He soon realized that, on account of being populated by a single, potentially non-native species, the soon-to-be-razed cedar stand at his temple was "artificial."

Hioki was also fascinated by the concept of *creating* a forest at all. Deciding to pursue the possibility, he contacted the book's author. "Professor Miyawaki called me back right away," Hioki said, smiling as he recalled the botanist's determined personality. Miyawaki gladly accepted the monk's proposal and visited to study the land around the temple in 2006. The following year, he led the first of five annual events to plant a total of 32,000 seedlings from sixty native species. Thus began a friendship and alliance based on a shared reverence for nature and awareness of humanity's dependence on it.

When the earthquake and tsunami hit three years after the temple forest was completed, Miyawaki called his friend to see that he was all right. He also wanted to discuss an idea that was lighting up his imagination, about how to help the stricken area recover. Miyawaki's proposal was to plant a winding 300 km (186 mi.) long, 100 m (328 ft.) wide native forest wall all the way up the coast that could slow down and reduce the strength of a future tsunami, affording people precious time to evacuate.

"A month after the tsunami, he came to the city and we started to research the coastal area, the devastated area," Hioki said. They observed that the majority of pine trees had fallen. Beginning 400 years ago, Hioki explained, fast-growing pine trees were planted along the coast to protect farms from sand and salty wind. Pines are not native to coastal areas, but they grow easily in sand. People are very fond of the pines-on-white-sand landscape, which has become emblematic of the area.

"Because their roots are not so deep, it's very easy for them to be washed away," Hioki said. Once uprooted, the floating pines then became dangerous ramming objects in the turbulent water. "It was shocking," he recalled of the destruction. By contrast, he observed that large, deep-rooted native trees had survived, along with many smaller native trees sheltered in the space around the big ones. Hioki's observations confirmed what he had learned from Miyawaki: Native forests have a natural resistance to stress that artificial forests do not.

Yet, despite their inherent fortitude, native forests are rare in the coastal area, according to Hioki. The emblematic pines, as opposed to native forests, had become a traditional feature of the landscape. And herein lay a significant obstacle when Hioki and Miyawaki presented their native forest seawall proposal to mayors up and down the northeastern coast. Many cities did decide to replant forests, but they chose the classical pine plantation instead of grappling with the unfamiliar "potential natural vegetation" approach.

"I sent letters to the mayors throughout the coastal area," Hioki said. "Only the mayor of Iwanuma accepted this idea, and it was because he had previously listened to a speech by Professor Miyawaki." Hioki explained that Miyawaki's method and theory were too new and difficult to understand, especially in a context of shock over the loss of almost everything that had been familiar. The Iwanuma mayor was also familiar with the Miyawaki mini-forest at Hioki's Rinnoji Temple in neighboring Sendai. Thus, he was prepared to absorb the philosophy and to understand how a diverse, native forest could better protect people in future disasters.

"The Iwanuma mayor had been to my temple for a tea ceremony, so he knew both Dr. Miyawaki and me." The three men discussed the project proposal. Hioki chuckled as he recalled how Miyawaki almost would not take no for an answer. Hardly able to back out in the end, the mayor promised to accept the project.

The first Millennium Hope Hills planting event in 2013 drew 4,000 volunteers, including local schoolkids, foreign students, Sango employees, and celebrities, who attracted media attention, to plant 30,000 trees. Over the next several years, thousands of volunteers planted a grand total of 350,000 trees and shrubs, with the last batch taking root in June 2021. The seeds of twenty-five local species were collected locally, and seedlings were grown on site in greenhouses managed by Espec Mic.

Hioki's Forest Seawall Association (FSA) planted an additional half dozen smaller seawall forests as opportunities arose in towns and cities as far south as Nagoya. Among these were two initiatives to bring life back to the coastal Arahama district of Sendai City. In 2016, the association responded to a call for proposals from the national forest agency. The following year, they planted a 1 ac. (.4 ha) strip of native forest as part of a 6 ac. (2.4 ha) project, which was dominated by 8,000-plus black pines planted by the other participating groups. Between 2012 and 2018, the forest agency and partnering companies and civic organizations planted more than 100,000 black pines, plus a tiny proportion of other species, including those in FSA's miniature Miyawaki forest, along the coast of Sendai City and two neighboring towns.

The FSA also contributed to Sendai City's Hometown Forest Regeneration Project in 2018. Here, the team planted a few dozen circular Miyawaki mini-forests measuring 4, 8, and 12 m wide (13, 26, and 39 ft.). Takano was pleasantly intrigued to discover that after three years, the smallest plots of barely more than 12 m^2 (129 $ft.^2$) were growing as well as any bigger Miyawaki forest. The geometry of these round patches may explain part of their success. Inspired by penguin flocks huddling in circles against arctic winters, the round design is meant to protect trees from rough coastal winds by maximizing the amount of sheltered internal space relative to the exposed edge.

These islands of trees were also symbolic of a small offshore archipelago that shielded Matsushima, a coastal town just north of Sendai, from the tsunami. "Although the entire Matsushima town was also flooded to the depth of about one meter, it avoided being physically destroyed by the strong waves," Takano explained. The different-sized islands in Matsushima Bay broke and dampened the forces of the tsunami. "We mimicked these islands of Matsushima by planting in a scattered polka-dot style," Takano said.

Seen from above, five of the round patches are set apart and arranged in the familiar pattern of the Olympic rings. Takano added that this touch was a nod to the support given by the 2020 Tokyo Olympics for the recovery and reconstruction of the tsunami-affected Tohoku region.

With the FSA's main project at Millennium Hope Hills complete, the association's focus today is simply to take care of the park, welcome its visitors, and teach as many as possible about the value of native vegetation. "Our goal is to share the Miyawaki Method and potential natural vegetation theory in Japan and the whole world," Hioki said. "We are making a monoculture society with humans on top of the world, and that is a very dangerous way to live. We should change our minds. Tree planting supports society by making it more diverse."

Miyawaki's vision of a 300 km (186 mi.) native forest seawall seems unlikely to be realized. Eleven kilometers (6.8 mi.) at Millennium Hope Hills, plus a few other dispersed patches, are dwarfed by a 12.5 m (41 ft.) high concrete seawall that has since 2011 been built by the national government along hundreds of miles of coast, not to mention the replanted pine forests. Hioki says that while it is hard to know what type of infrastructure will best protect people in the future, it is critical to at least be able to test the potential natural vegetation theory so that it has the chance to become a valid, recognized option one day. "We have no results or evidence yet, so we should wait for results. In twenty to thirty years, the next generation will judge what is the best way to protect their lives."

Great Challenges at the Great Wall

"As you well know, the Great Wall of China is a World Heritage site," Miyawaki said in a 2012 interview.[7] Stretching clear across northern China from the Yellow Sea, east of Beijing, to the Gobi Desert in the west, the Great Wall is "an absolute masterpiece," according to UNESCO, "not only because of the ambitious character of the undertaking but also the perfection of its construction." Building started in the third century BC and continued over the next two millennia, resulting in "a perfect example of architecture integrated into the landscape."[8] Perfect but for the deforestation that paved the way for its construction.

The landscape that the wall traverses north of Beijing is mountainous. To fire the ovens that cooked the bricks to build the wall many hundreds of years ago, as Miyawaki went on to explain, trees along its path were cut and burned. "Therefore, all native forests have been destroyed around the area and it is semidesertified," Miyawaki said. "It doesn't rain much near the Great Wall, but when it does, topsoil runs down." In modern times, logging and grazing have prevented the forest from growing back.

By the time Goulin Xu arrived at the Great Wall in 1998 as a member of Miyawaki's planting team, the slopes were covered in grasses and shrub but lacked trees. "There was no plant community that could be called a forest," he wrote to me in his second language, Japanese, which was then translated into English for me. He added that he recalled seeing forestlike vegetation in a distant valley. Originally from northeast China, Xu lives in Japan, where he has worked for Espec Mic and practiced the Miyawaki Method since 1992.

Miyawaki had been invited in 1996 by the City of Beijing and its Japanese funding partner, Aeon Environmental Foundation, to restore the land around the Great Wall in a multiyear planting project. Xu joined the effort as both a technical staff member and interpreter. "When I heard about the Beijing project," Xu wrote to me, "I thought I would never be so happy as I would be if I became involved in the Miyawaki Method reforestation project in my home country."

"Having both language ability and specialized knowledge," Xu explained, "I think I was helpful for the communications between the Japanese side and the Beijing city government side. In particular, my role was to convey the core technology of Miyawaki-style forestation that was commissioned by the Beijing Municipal Government to the local staff in Yanqing County."

Previous reforestation efforts at this iconic site had failed, according to Miyawaki, due to the wrong choice of species. The botanist recalled the mayor of Beijing challenging him over dinner to plant trees that would not die soon after planting. "After three years, all that's left is the sign board, no forest," the mayor complained of prior reforestation attempts. "We want a genuine forest that will prevent yellow sand [from blowing into Beijing], store water, and function as watershed protection." Most of Beijing's drinking water comes from the mountainous Miyun Reservoir watershed, through which the Great Wall passes north of the city.[9] "Can you do this, professor?" the mayor asked him.

"I said to him, 'that's why I'm here.' Until then, they had been planting the so-called 'fast-growing trees.' In other words, fake trees such as willows, alders, false acacia [black locust], and poplars," Miyawaki explained in his interview. "These plants grow fast, but they don't last long. Real trees endure all conditions."[10] Restoration efforts such as the Grain to Green Program in China have relied on fast-growing trees to stop soil erosion and reduce wind speed. However, the species chosen are not necessarily best suited to the local environment, nor planted in diverse mixtures.[11]

The mayor asked Miyawaki what he was going to plant. "I had found out the answer by then," Miyawaki recounted, "so I said '*Quercus mongolica* [Mongolian oak].'" Identifying this native species had not been easy: "Searching for potential natural vegetation is like looking at something hidden beneath a thick cover," Miyawaki said. He, Kazue Fujiwara (plant ecologist and former student of Miyawaki), and a few of their Chinese colleagues had conducted field surveys of the surrounding area, including at a national forest reserve not more than an hour by car northwest of the planting site. They found that Mongolian oak grew both in the protected forest and in a spot near the wall that wasn't grazed by sheep or goats.

Fujiwara drew up a species list that included, in addition to oak, Chinese red pine (*Pinus tabuliformis*), painted maple (*Acer pictum*), slow-growing Chinese thuja (*Platycladus orientalis*)—a conical cypress tree with flat, scaly, needlelike leaves—and the small-statured smoke tree (*Cotinus coggygria*), whose flowers grow into purplish plumes that resemble smoke from a distance. The vast majority of what was planted was oak, though, and Xu was among those tasked with collecting and then germinating acorns in small pots in a greenhouse. Crews of mainly local people collected 1,000 acorns or more per year throughout the project.

Xu recalled how, leading up to the first planting in 1998, the organizers had to figure out how to adapt the Miyawaki Method to a variety of local challenges: "It was impossible to carry out the Miyawaki Method in exactly the same way as we do in Japan." The ground was rocky and lacked topsoil and, on account of the site's remoteness, they did not have access to compost, rice straw (mulch), or running water. Annual rainfall is scanty in the area, with as little

200 mm (8 in.) falling in some years, according to Aeon records, and only in summer. Winters are completely dry, windy, and very cold, with temperatures dropping to -20°C (-4°F).

In the absence of the aerated, compost-enriched mounds that Miyawaki's teams typically create to ensure a strong start for saplings at smaller and more accessible sites, the young trees were planted directly into 60 cm (24 in.) holes dug with shovels.[12] Xu explained that they cut the grasses and shrubs growing at the planting site to use for mulch instead of using straw mulch, which was not available locally.

The key was to dig deep enough to enable the roots to penetrate the soil as quickly and easily as possible. Learning as they went along, Miyawaki observed afterward that: "If we don't dig deep enough, trees don't grow well. Areas where the digging was done sufficiently have been successful."[13] In those successful areas, Miyawaki said, trees had grown to 3 to 5 m (10 to 16 ft.) or more by 2012, fourteen years after the first of three plantings.

Xu continued to staff the project over its first three years (1998 to 2000), during which time a total of 7,400 Japanese and Chinese volunteers and workers from the Beijing Yanqing Prefectural Greening Commission together planted 390,000 saplings on an area of about 30 ha (74 ac.). "I stayed there for about a week when preparing for the tree-planting festivals," Xu said. "From the second time onward, as we got on the right track, I think I went to the site about four times a year, including during the festivals."

The planting festivals were an important aspect of the project, which from its conception at the 1995 "2nd International Symposium on Environmental Issues between Japan and China" emphasized binational collaboration and friendship. "The significance of this project is also great in the sense that the message on the importance of nature was transmitted from the Great Wall of China, which is designated as a World Heritage Site," states an excerpt from Aeon Environmental Foundation's thirty-year history.[14] Volunteers were transported to the site in more than fifty buses. They gathered for a formal welcome and introduction to the planting method, and then divided into 100 groups of twenty to thirty people each to plant the trees.

When Aeon staff surveyed the planting area in 2002, they documented a 70 percent survival rate. However, 70 percent is a decent outcome given the site's harsh conditions—a strong majority of the trees survived. Even so, Aeon decided that additional plantings and more maintenance were needed to establish the young forest. They organized two additional three-year planting projects, this time with smaller volunteer teams that first replaced the saplings that had died and then expanded the total reforested area. This time, they installed water tanks on site to be able to water the saplings periodically.

Fujiwara visited the site in July 2006 with Box and some of Fujiwara's graduate students. A picture of that visit depicts the small research team and their local guides standing in a hilly green expanse of healthy shoulder-height oaks, a few pines, and what appears to be a young smoke tree in the foreground. The Great Wall is just barely visible along the ridge on the horizon. The distant mountains are speckled with bare tan patches, which Fujiwara explained is what the oak-covered foreground looked like prior to the start of planting in 1998.

The last time Xu returned to the site was in 2016 or 2017. He reported that the whole landscape had greened up due to a variety of restoration efforts, including other plantings and restrictions on grazing and firewood cutting, but that the Miyawaki Method site stood out due to its distinct vegetation mix.

The Great Wall project is compelling not only for its emblematic location and massive volunteer effort, but also due to the giant challenges that needed to be overcome. Navigating linguistic and cultural differences, the binational team brought back a nearly lost local species from seed and made it grow again on a steep, degraded, dry site not easily accessed for maintenance. Moreover, the project organizers accomplished this with broad public participation in order to impart a message and spirit of cross-border collaboration for healing the earth.

Xu appreciates "the degree of technical perfection" called for by the Miyawaki Method, which he said is "backed by the policy to 'not cut corners, and not let others cut corners.'" It seems that high standards are even more important when big obstacles threaten failure. In the end, the effort Xu and others put toward meeting those standards was worth it. "When I look up at the forest that grew up in this way, I am proud of how happy Dr. Miyawaki would be if he could see this forest."

CHAPTER SIX

Healing Forests

The forest is the root of all life; it is the womb that revives our biological instincts, that deepens our intelligence and increases our sensitivity as human beings.

—Akira Miyawaki[1]

Adib Dada had seen Shubhendu Sharma's online talk when he decided to plant a Miyawaki mini-forest on the Beirut River in Lebanon. An architect trained in biomimicry (a design approach based on nature as the model), Dada's work aims to transform degraded urban land into public space in a city where parks are scarce. Beirut has an estimated one square meter per capita of green space.[2] In particular, his dream is to restore the cement-encased river.

The river was channelized in 1968 to mitigate flooding after development on the floodplains.[3] Before it was paved, Dada told me, explaining how his discovery of the river's demise ultimately led him to the Miyawaki Method, the river and its floodplains were a riparian ecosystem whose vegetated surfaces regulated temperatures and absorbed floodwaters. It was also a cultural site where festivals were held and had long served as a source of drinking water. Now it carries industrial waste and heaps of trash, and it has been abandoned by local communities who no longer see any reason to gather beside it. Dada said the paving of its bed amounted to the river's death.

Short of ripping out the concrete banks, Dada's vision for the Beirut River is to restore the native vegetation around it—from the mature natural forest in the upstream basin all the way through the city to the coast. He began to propose remedies in 2013, including the installation of vegetated swales (dips) along roadways to divert rainwater and let it infiltrate into the ground, and the creation of riverside parks that would become rain gardens, absorbing stormwater during the rainy season.

For some time, city leaders could not be bothered. But in 2019, Dada had a breakthrough. His friend Elise Van Middelem, knowing of Dada's interest in the environment, had contacted him with an idea. She, too, had recently discovered the Miyawaki Method via Sharma's work with Afforestt. In the past, Van Middelem's work as a creative brand strategist in the fashion, art, and tech fields had taken her to Beirut, where she had met and become friends with Dada. Now she was in the middle of switching gears professionally. Having recently moved back to her native Europe from California, Van Middelem was about to launch SUGi, her fundraising platform focusing specifically on biodiversity and ecosystem restoration.

Van Middelem's idea was to work with a friend on her first SUGi-orchestrated Miyawaki planting project. Dada did not hesitate to accept this invitation, which fit his vision for river restoration. Van Middelem organized a conference call with Dada and Sharma, the three hit it off, and a plan was made.

With partial funding from SUGi, Dada hosted a weeklong May 2019 workshop attended by fifteen local volunteers who were selected from sixty applicants. Sharma led the workshop, training participants in the Miyawaki Method. The week started with the study of native vegetation and lessons on soil health and biodiversity and finished with the forest planting. More than 100 people pitched in to help plant 200 m^2 (2,153 ft.^{2}) along the Beirut River with native forest species, including Palestine oak, fragrant Greek sage, and the small-statured Judas tree with its heart-shaped leaves and pink blossoms in spring. They also planted what has become Dada's favorite, the strawberry tree, which bears sweet red edible fruit and has thin, deep red bark that is layered over pistachio green beneath. Six months later, they planted the same species mix on an additional 300 m^2 (3,229 $ft.^2$) at the same site, naming the finished grove Beirut's RiverLESS Forest.

"People have flocked to this project," Dada said. "You can feel the hunger for it." People who happened to walk by the site on planting days stopped to put a sapling in the ground, some saying they had never really touched a plant before. Other volunteers had helped to prepare the site by picking up trash and sorting it for recycling. A few serious volunteers have taken on regular watering and weeding as the plot becomes established.

In the short time since it was launched, people have flocked to SUGi, too. From the United States to Kenya to Chile, Van Middelem has fielded requests for assistance in the Miyawaki Method. The SUGi team connects people to trainers like Sharma and helps with fundraising and project communication. Van Middelem refers to Dada and the other "forest makers" in her quickly growing international network as members of the "rewilding generation." Scientists and environmentalists use the term *rewilding* to refer to a restoration approach that privileges natural ecosystem processes—such as pollination by insects, seed dispersal by various animals, and control of the defoliation effect of herbivores by top predators—as a cure for degradation. The ultimate goal of rewilding is to eliminate the need for human intervention in an ecosystem because it has become self-sustaining.

The Miyawaki Method takes a rewilding approach to afforestation because it aims to re-create natural forest communities composed of plants that have coexisted and interacted in a particular place for millennia. The reintroduction of the native plant community in turn serves as an open invitation to native animals and microorganisms to return and complete the local food web.

One afternoon in 2020, about a year after Dada and team planted Beirut's RiverLESS Forest, Dada had just left the office of his Beirut architecture firm and was in the underground parking lot when a stockpile of ammonium nitrate exploded at the port a mile away. The blast shattered the windows of his office and trapped him under a gypsum ceiling, which fell in a sheet on his back as he leaned over his toddler to shelter her from the falling rubble. The pair was stuck for about twenty-five minutes before parking attendants managed to clear open a space for Dada's child to emerge; he then slid out on his back from underneath the pile.

Dada was badly hurt, but he could walk. He eventually made his way to a hospital, where he was put in a back brace. He was among at least 6,500 people to have been injured in the blast, which was not intentional, but rather the outcome of government negligence. More than 200 people died and 300,000 lost their homes; others lost their businesses. Thoughts of COVID-19 faded far into the background as people poured into the hospitals that remained standing; some hospitals had been destroyed by the blast.

Fluffy toxic clouds lingered for days, Dada said, and the ground was covered in 800,000 tons of rubble, which the UN reported was likely to contain hazardous chemicals.[4] "It's been so emotionally draining," he said. People are traumatized, he added, even those who were not directly affected by the blast, but had family members who were. Several months later, loud noises still triggered feelings of trauma for many. "Everyone feels violated," he said. A year later, nobody had yet been held accountable.

As he himself recovered physically, Dada thought about how to help others put their lives back together. "I really believe in nature healing people," he said. Thus, he came up with the idea to plant a forest as a "living memorial" in a neighborhood affected by the blast. But rather than dwelling on the past, a native forest memorial would look to the future. "We can heal the land damaged by the explosion and urbanization," Dada said, "while the land heals us through the act of planting."

His original plan was to plant a large area of at least 400 to 500 m^2 (4,300 to 5,400 $ft.^2$) near the epicenter of the blast, and to collaborate with mental health providers to facilitate nature-based therapy for people suffering from trauma. By planting and then tending the trees over the couple of years it takes a Miyawaki mini-forest to become self-sufficient, the participating community would simultaneously strengthen their own inner resources.

Because the explosion site at the port is in the city center, where land is expensive, no terrain of this size became available. Instead, Dada found a public park that appeared to have been abandoned in a neighborhood a mile from the epicenter—it contained dead trees, shrubs, and grass; three broken water fountains; and lots of litter. Collaborating with two other socially minded groups who installed a playground and handicap access, Dada's team planted two very small forest patches, each less than 100 m^2 (1,076 $ft.^2$)—the equivalent of about six to eight parking spaces.

On August 4, 2021, marking the one-year anniversary of the port explosion, Dada and fellow volunteers, including a couple of families who had lost loved ones a year before to the day, planted 254 saplings in this park. Each sapling represented a victim of the blast and was planted with a slip of handmade paper on which the deceased's name was handwritten. It was on this planting day that Dada unexpectedly learned of the passing of his inspiration, Dr. Akira Miyawaki, in July 2021, following a long illness brought on by a stroke. Dada's team instantly decided that one of the groves would be dedicated to the visionary man whose urban eco-restoration technique they had inherited.

Both memorial plots would be watered and weeded by a few of the local families who had someone with a name buried in the soil among the trees. A second team—a group of at-risk children and their counselors in a play- and nature-based therapy program—would care for the baby forests on a regular weekly schedule.

These memorial forests, Beirut's RiverLESS Forest, and several other proposed and finished projects are all a part of an initiative, organized by Dada's architecture firm, called theOtherForest, whose goal is urban ecological and social regeneration. theOtherForest's approach to social change is well considered, its choice of projects thought-provoking. "For us, the projects are always a very tangible way to start a conversation," Dada explained. "Because from our experience studying the Beirut River for six years and doing talks and conferences and lectures and community meetings and exhibitions, all of that was too abstract and didn't lead us anywhere. So that's why we decided to plant a forest."

While the memorial forest creates a therapeutic, living sanctuary that pays homage to the victims and survivors of recent tragedy, Beirut's RiverLESS Forest beckons people to the banks of a dead river. The unexpected presence of a lush grove is an impetus to get people "looking over that concrete wall, discovering the river of sewage, asking questions, and then hopefully deciding to do something about it," Dada said.

The Power Plant Forest initiative in Zouk Mosbeh, a town just north of Beirut, extends a similar invitation to observe the dystopian reaches of overurbanization. In collaboration with the municipality and a local activist, theOtherForest located 180 m^2 (1,938 $ft.^2$) wedged in the center of a three-way intersection in which to plant a seventeen-species native forest. Behind this plot, a notorious power plant continuously dusts the city in oily soot, a contributor to a grim statistic: Lebanon has the highest premature death rate from fossil fuel pollution in all of the Middle East and North Africa.[5] The hope is that this mini-forest will draw attention to air pollution in the way the RiverLESS forest draws attention to the trash-filled river.

In the midst of spiraling economic and political crises, Beirut was running out of fuel by August 2021. This threatened the municipal water supply and caused long blackouts, both of which contributed to extreme hardship for the public.[6] "The price of water has tripled by now and the trend is upward," Dada said. "Everyone is immersed in day-to-day or hour-by-hour issues that are coming up—it's very surreal what's going on."

Lack of fuel immobilized city water trucks that would otherwise irrigate the still-young Miyawaki groves around town. Dada's team had once paid a premium to send a private truck to water the forests, but this was not a lasting solution, and without a residential area nearby, no hyper-local volunteer could be called upon to haul jugs of water to the site. Coping as best they could, the team held their breath for the rainy season to begin in late fall.

Dada's team also decided to delay new projects for late fall of 2021, to sync with the seasonal rain cycle. One planned forest would be on the land adjoining Beirut's RiverLESS Forest, where a few exotic eucalyptus trees had been sucking up more than their share of water and nutrients, and spreading quickly. "We had discovered it was a common practice in the French colonies to plant eucalyptus to dry out [what were unfavorably considered] swamps. So, in Africa, they have this invasion of eucalyptus trees," Dada said, explaining that eucalyptus roots reach deep within the soil in search of water, easily drying out the ground.

"This is something we noticed all over Lebanon as well," he continued. "And the fact is that this species outside of their native habitat are very invasive because they don't have the animals that can eat them or the insects that can decompose their leaf matter. So they also become a fire hazard because there's a lot of leaf litter on the ground and it's very oily." With financial support from the Embassy of Switzerland, Dada's team planted 2,400 native tree saplings in November 2021, replacing six eucalyptus trees and more than doubling the size of Beirut's RiverLESS Forest.

Three additional 200 m² (2,153 ft.²) public forests were slated to be planted, one per year, between 2021 and 2023. While Dada would have preferred to consider potential planting sites in dense urban areas throughout Lebanon, such as in Tripoli, the economic situation limited the options to Beirut for the time being. The team scouted for places where a tiny pair of green lungs could have the biggest impact. "Eventually, we'd love to have a map with different layers," Dada said, explaining a way to systematize the search for high-impact mini-forest sites. Such a map would indicate areas of the city with high flood risk, high temperatures due to building density, and a lack of greenspace. Overlapping burdens would signal priority planting sites.

I asked Dada if he thought the momentum of theOtherForest's first five public projects, all planted in about two years, could be opening space for a public conversation about his bigger vision for Beirut River restoration. Dada had told me that the river valley sits along an important bird migration route, making the habitat it provides critical.

"Honestly, it's been such a difficult year," he said, thinking back on 2020. "You know, if it wasn't a priority before, now . . . people are really scrambling just to survive. More than half of the population is under the extreme poverty line."

He paused. "We know that the work that we're doing eventually will have a positive impact," he added, reflecting on Lebanon's air pollution crisis, "but it's difficult to explain this to people who are now very much under stress."

No explanations were needed for a small group of active volunteers, though. Dada has seen a handful of local people, who joined their first public forest planting out of curiosity, become informal spokespersons for native biodiversity. These folks now stop by the young forests for impromptu weeding or watering, document wildlife sightings, and suggest dilapidated spaces as future planting sites. Others who may have volunteered just once are visibly transformed within the passage of a few hours—especially the prim and tidy professionals who temporarily become unashamedly smudged and sweaty earth-tenders, completely absorbed in the soil.

Back to the Roots

It took about five days to prepare the soil and plant 2,500 native trees onto 700 m^2 (7,535 $ft.^2$) of barren ground outside of the Yakama Nation Correctional and Rehabilitation Facility in Washington State—the first of two sections of the Miyawaki-style Healing Forest. Planting until sundown amid a light snowfall, the correctional facility team and a handful of other community members organized the saplings into the shape of a medicine wheel, a traditional Native American design associated with health and healing, where spokes mark the cardinal directions. The Healing Forest is designed with four paths leading into a large central talking circle.

"You know, I'm going to be able to come by here and show my kids that I did that," says a tree planter in a sanitary mask and orange jumpsuit. He was recorded in a short video of the planting day created by SUGi, which helped fund the effort.[7] "It feels good. It makes my heart feel good."

For Chief of Corrections Vernon Alvarez, the Healing Forest is a creative approach to rehabilitating people struggling with problems like drug and alcohol addiction. The mini-forest serves as an outdoor classroom, where tribal elders can teach traditional skills and case workers can meet with their clients. When sessions are held in this natural setting, the fresh air, sunshine, and bird sounds open people's senses and minds. Alvarez says the idea is to make an impression, to create a significant experience for people who are working through major life changes. The goal is also to connect these folks to their heritage as members of the Yakama Nation.

"We found out a lot of the inmates here had [traditional cultural] teachings from their grandparents, their extended family, and of course from their mom and dad. Well, as they were growing up, they lost touch with that," Alvarez explained to me from his desk as we spoke by video.

The cultural aspect of the rehabilitation program is broad. Residents learn to grow their own corn, tomatoes, and herbs, which they then cook into stew with wild game like salmon, deer, and elk. They also learn traditional dancing and drum making. Until COVID-19 struck, they took part in sweat-lodge sessions.

Elders like Marylee Jones teach residents about native plants—their Yakama names and many traditional uses. For example, red osier dogwood (*Cornus sericea*)—a flowering shrub with red branches—can be carved into a digging tool called a *kupin*. This tool is used to gather bitterroot in the foothills of nearby Mount Adams, also known as Pahto, at the western border of the Yakama Reservation. Gathering wild roots is a passion for Jones, who takes small groups from the facility to Pahto in spring, summer, and fall, teaching them how to identify traditional food plants and gather them in a way that respects the plants as sisters. After all, these plants have been with the people of this land from time immemorial.

Jones shared with me the Yakama creation story: In the beginning, the Creator made plants and animals. Then the Creator made people, who were naked and vulnerable and unable to survive on their own. "What am I going to do with you?" groaned the Creator. The plants and animals, who had been listening, responded by offering themselves to feed, clothe, and heal people when they became sick.

"In return," Jones concluded, "we promised to take care of them and preserve their habitat—the water, soil, air." She explained that understanding this relationship with wildlife allows people to honor plants and animals as family members. A plant is "not just a landscaping item, not just a weed," she said. "It's something with much deeper value."

"There is an instant connection with the plants," Jones said, describing residents' visits to the foothills. While gathering and later washing the roots, one must show respect for the plants through personal cleanliness, sobriety, and a present and positive state of mind. "People discipline themselves to be able to respect the plants," Jones said. Once the roots are washed, the residents then share them with other members of the tribe.

This ancient tradition has nearly died out on the reservation, although Jones herself was born into gathering as a way of life. Her earliest memories are of being bundled up in the back of the car before daybreak and waking up at dawn at the edge of the mountains with her grandma, mom, and aunts.

"Good morning, did you sleep good?" her grandma would ask before tying a basket around the young Jones's waist. Grandma would dig and chat with her little apprentice, while Jones chatted back, observing the elder's work. "Seeing what she was doing, I would do it. She didn't ask me to hurry. The whole thing was like entering into a beautiful space."

Before there were highways, stores, and bank accounts, Jones explained, gathering was a way of life for all Yakama women—it was how they provided for themselves and their families. By 2011, Jones and her mom could find only about ten other women to go out gathering with them. Having recently returned home after some time living in Oregon, Jones was shocked to discover how this tradition had all but disappeared among her people. Similarly, she said, very few still speak the native language, and young people are easily distracted by things like designer shoes.

"We are so acclimated to the mainstream way of life," she explained, "that with every check cashed, our culture and identity slip further away—to the point where they are no longer considered important." The situation has reached the point of emergency in Jones's view. And it's this loss of cultural identity that underlies the behavioral problems that land some in the correctional facility.

Jones teaches inmates about intergenerational trauma, which exists in her own family. In the late 1800s, she said, her great-grandfather was a child. One day, he and some other kids saw a wagon passing by while they were out playing in the valley. It was waving something funny behind it that looked like a pair of pants on a stick, and the curious children approached. They were snatched up by the riders and taken to Fort Simcoe Indian School in White Swan, Washington, never to return home again as children.

For nearly 100 years starting in the 1870s, many native children around the country were taken from their families and forced to attend boarding schools.[8] Their hair was cut, their names were changed, and they were dressed in Western clothing and forbidden from speaking their own languages. The US government's goal was to assimilate native children into the settler society. "His memory was a wipe-it board," said Jones, explaining that everything her great-grandfather had known, including ways of behaving, was purposefully erased.

The forced removal of children from their families and other atrocities, ranging from physical violence to the prohibition of traditional Yakama songs and language, created a generalized sense of distrust within the tribe. Jones described how distrust settled into the community "like a fog," remaining even today.

But the inmates and others outside the facility are becoming receptive to the old ways and to plant-centered traditions. Jones started announcing on social media whenever she and her mom were going gathering, promising to initiate newcomers on how it is done. Little by little, the community of gatherers expanded until about 100 women and girls had taken up the practice.

Once beginners get started digging, "it's like the blood in their bodies starts to bubble like 7UP," Jones said. "All the ancestors are released. It's beautiful when you see the children finding roots that people would normally think of as weeds."

Some of the sacred roots found in the nearby forests have been planted at the edge of both the Healing Forest and another mini-forest not far from the correctional facility, in the backyard of dietitian and master gardener Debra Byrd. Like Jones, Byrd is excited to start teaching more people about the tribe's sacred plants. In fact, it was Byrd who first proposed a Miyawaki mini-forest for the reservation.

In her role with the reservation's diabetes prevention program, Byrd met fellow dietician Mary Baechler, the aunt of Seattle-based architect and Miyawaki-method practitioner Ethan Bryson. When Bryson's forest-planting ventures cropped up in conversation between the two women, Byrd invited him to visit and discuss a potential mini-forest on the reservation.

When Bryson met with Byrd and Alvarez, they discussed two forest ideas—the Healing Forest at the correctional facility and another behind Byrd's house. Byrd wanted to create a peaceful, spiritually meaningful place where she could teach youth about native plants and their uses. She, too, sees the younger generation losing touch with food traditions, so she plans to host Head Start, a public preschool program, and school field trips at Tu'paxin, the little forest in her backyard.

Tu'paxin is Byrd's tribal name. She was named after her great-grandmother, whom she told me over the phone was a strong, protective woman who knew all the plants and medicines. Some twenty-five people came to help plant Tu'paxin.

Both forest plantings began with a blessing ceremony led by Jones. "Today, I'm really happy for all of us to be here to celebrate this beautiful part of where we're at in our lives," Jones said. "By doing this today," she continued, her voice quavering with emotion, "we have a place right here where these beautiful sisters are going to grow."[9]

Bryson recalled watching people's demeanors change as they engaged in the ceremony and subsequent plantings. Some of the residents looked at first as if they wondered what the heck they were doing digging holes in the ground, but finished the day full of energy and determination.

"Excitement happens," Bryson said. "You could see it in their eyes. Some almost couldn't stop planting." Fortunately, the chance to do it again came in the spring of 2021, when the Healing Forest doubled in size with the planting of another 2,500 trees. Even residents who had been released since the first planting in October wanted to come back—to have an ongoing role in caring for the young forest.

As the trees take root, the forest's presence is helping to root the inmate-caretakers into their own heritage. "Now they're aware of their own culture, their own curriculum, their own calendar," Jones said. "The forest will teach every one of those inmates that a swallow coming by means the salmon are running," she said—an important signal for a people whose traditional diet is based on salmon. "The plants are coming up out of the ground," Jones concluded. "We're also coming up."

Growing Gratitude

In late November 2020, a masked and socially distanced team of landscaping professionals and conservation volunteers planted a 1 ha (2.5 ac.) Miyawaki mini-forest in an East London city park. Local residents and families would have been there too, aiding in the enormous task of planting 28,000 saplings in less than a week, if not for the restrictions imposed in response to the COVID-19 pandemic.

On the other hand, this Forest of Thanks was conceived of only *because* of the pandemic and the accompanying sense of urgency to recognize the members of a community slogging through it. “We needed to have something we all agree we can celebrate—a place we can go, like a bench, a carving—but something more than that,” said Barking and Dagenham Council leader Darren Rodwell, who first proposed the idea of a forest.

“I wanted to make the Forest of Thanks as a way to thank the people that have kept us safe, supported our loved ones,” Rodwell explained. “We wanted to thank the residents for all the work they’ve been doing to keep their families safe. We wanted to honor the people who have passed, and to show how thankful we are for those who are still here.”

Barking and Dagenham was hit particularly hard by COVID-19. By mid-April 2020, the borough had already lost 100 people. By January 2021, it had the highest infection rate in all of greater London, which itself had a rate higher than the national average.[10] The more contagious variant was circulating, and a local hospital had already registered 1,100 dead from the virus.[11]

The reason for this working-class town’s pronounced struggle is a familiar one, recognizable everywhere in the world: Poverty worsens health outcomes. More than a quarter of Barking and Dagenham residents, including half of its children, are considered to be in poverty.[12] The town’s unemployment rate is high as well, while those who are employed constitute a high proportion of essential workers.

“Because we are the poorest borough, the jobs people do are not fancy, but they are essential: cleaners, drivers, security guards . . . all jobs where people need to work outside the home. More people have manual jobs here than elsewhere in London; we have a lot of working poor,” Rodwell explained.

At one time, he said, the city was a blue-collar industrial powerhouse. It supplied England with cars, telephone cables, precast concrete panels, and pharmaceuticals, among other goods. In the late 1920s, Ford Motor Company built its biggest European factory to date on the Dagenham banks of the Thames River, where by the 1950s it employed as many as 50,000 workers. The London suburb remained a manufacturing hub for several decades.

Various plants began to shutter after the 1980s, though. In 2002, the Ford plant shut down most of its operations, leaving open only a small engine manufacturing division. Along with their final paychecks, workers were left with respiratory troubles associated with asbestos exposure and the generally dirty, dusty air of the smokestack-filled area.

In the physical and economic vacuum of departing companies, Barking and Dagenham is shaping itself a new identity. As it works out how to leverage empty industrial sites toward the goal of creating 20,000 new jobs, the city council envisions a "clean and green" twenty-first century, Rodwell said. Switching gears with a modern portfolio, the council's plans include a solar farm, wind turbines, a data center that recycles heat, a film studio, and a science and engineering facility.

"We want to be the green capital of the capital," Rodwell said. To this end, the council has declared a climate emergency, aiming to become carbon neutral by 2030. Planting trees is part of the strategy—in the city's large stretches of public green space, which are largely covered with mowed grass, and in people's yards where the residents can be incentivized. The Forest of Thanks grows in this context of an environmental remaking too, making it a living juxtaposition of two global crises.

Once approved by the city council in the spring of 2020, the forest project was assigned to Gareth Winn, ranger services team leader. Design, planning, and implementation were up to him. One thing he knew clearly from the start was that whatever was to be planted, it had to be big enough to express the depth of the community's gratitude for the frontline workers who had risked their lives daily. "A small forest would just get lost," Winn said.

Prior to 2020, Winn had never heard of the Miyawaki Method. He first encountered it in February, when Elise Van Middelem of

SUGi visited the borough to propose a 100 m^2 (1,076 $ft.^2$) Miyawaki mini-forest. With funding already secured for this forest and the novelty that it would be London's first, the offer was too good for the city council to refuse.

It was by a chance personal connection with a London arts and culture organization that Van Middelem landed in Barking and Dagenham. She had previously made the same proposal to a dozen London organizations, to no avail. But the East London borough was ready to go—that is, until March 2020, when COVID-19 protocols shut down all activities.

Communication between Van Middelem and Winn lapsed until early summer 2020, when Winn began assembling a project team for the Forest of Thanks. He contacted Van Middelem to say the project was up and running again, but that the forest was going to be 100 times bigger than originally envisioned. In addition to funding, SUGi contributed expertise in the form of James Godfrey-Faussett, a professional landscaper and lead forest maker for SUGi. He would train Winn's team in the Miyawaki Method and assemble supplies, including the seedlings, wool compost (actually made from sheep wool), and 24 tons of straw to properly mulch a hectare (2.5 ac.).

Winn also recruited members of the Thames Chase Community Forest, an initiative to regenerate 40 m^2 (104 km^2) of degraded land in Barking and Dagenham and neighboring towns, and the Conservation Volunteers, a national ecosystem restoration network. Getting the team together and combining everyone's unique skills was an invigorating experience, Winn said.

The urban ranger had initially been skeptical about the Miyawaki Method—at first glance, there didn't appear to be anything special about it: preparing the soil, planting trees—lots of people do that. But as he learned more, the method's unique features—the creation of conditions conducive to rapid growth and minimal maintenance—became clear. "And it's all natural," he added. "There's nothing artificial involved—no tree tubes, no plastic mulch mat, no synthetic fertilizers. And within three years it will become a mini-forest." Winn sees the method as a good recipe for urban ecosystem restoration. "What I'm hoping is that the project will be a catalyst for change, and people will be inspired about this new way of planting," he said. "If you can find enough small areas of green space to plant into, this can be hugely beneficial to air quality."

The site chosen for the Forest of Thanks is the northwest corner of the gigantic Parsloe Park, situated in between Barking and Dagenham. Central placement was critical, Winn said, to expressing gratitude evenly across the merged community, and to ensuring the forest's universal accessibility.

Surrounded by residential neighborhoods, Parsloe Park is a vast, flat, grassy area. Its defining features—a pond, a playground, basketball and tennis courts, a community center, and a handful of dispersed trees—are small in comparison to the park's about 58 ha (143 ac. or so) of lawn. Winn is looking forward to seeing the Forest of Thanks, planted in the shape of an oak leaf to symbolize strength and resistance, become a main feature of the park as it grows up in the next few years.

The Forest of Thanks project forged a partnership between SUGi and borough leadership that continued into 2021, when Godfrey-Faussett and the ranger services planted two more mini-forests—one at a school, another beside a busy road—both to mitigate air pollution. The team hopes to create Miyawaki pocket forests at six additional Barking and Dagenham schools in 2022, Godfrey-Faussett said.

The East London borough is not the sole community in the United Kingdom embracing the Miyawaki Method. After the town of Witney planted the nation's first mini-forest in March 2020 in collaboration with Earthwatch Europe, adopting the Tiny Forest program model of IVN Nature Education, more than fifty additional mini-forests had been planted by the end of 2021 at schools, parks, and elsewhere, from Cornwall to Glasgow.

CHAPTER SEVEN

Earth's Living Tissue

The focus truly needs to be on preserving the trees and the forest that we have first.

—Holly Paar, Dogwood Alliance[1]

Miyawaki was fond of the tabunoki tree (*Machilus thunbergii*), a long-living evergreen laurel. "I'm truly happy I got to see this tabunoki again after sixty-five years," he said with a glowing smile in a video recording, patting the trunk of a tree that survived the atomic bombing of Hiroshima. He had first visited this tree as a college student, shortly after the war, when red buds were sprouting up from its remains. "So happy I want to hold it."[2]

Perhaps Miyawaki appreciated the tabunoki's broad sheltering canopy, or its small, waxy, yellow-white, star-like flowers and sturdy glossy leaves that stay green in winter. However, it is probably the tabunoki's structural role in the native forests of southern coastal Japan that really earned this species a cherished place in the plant ecologist's heart.

Every natural forest type has its dominant species, which Miyawaki referred to as leaders, and which constitute the largest portion of trees he planted in his forests. These species are abundant, "grow tallest, and produce the greatest cover," influencing environmental conditions, plant community structure, and ecosystem function.[3] They are considered so defining that the forest types in which they grow are often named after them—for example, the beech-maple and oak-hickory forests of the eastern United States.

Along with the castanopsis (*Castanopsis cuspidata*), a cousin of beech and oak, tabunoki laurel is the leader of many native forests in Miyawaki's homeland—it is hard to imagine such forests without these species. Together with the smaller and less common plants, tabunoki and castanopsis form an integral biological community adapted to local conditions.

Elaborating on the ecological roles of biological communities in *The Healing Power of Forests*, Miyawaki wrote: "Creatures interact with the physical environment through a division of labor: some creatures are producers (green plants), some are consumers (herbivores and carnivores), and others are decomposers [fungi and earthworms, for example]. Taken together, all these interactions create a system of material and functional balance—the ecosystem."[4]

This term's etymology further illustrates its meaning. The prefix *eco-* is derived from the Greek word for household, *oikos*. Ernst Haeckel, the nineteenth-century German zoologist who coined the term *ecology*, wrote of "the place each organism takes in the household of nature, in the economy of all nature."[5] The postfix *-system* refers to interacting components that form a whole greater than the sum of its parts—an entity that exhibits its own unique qualities and functions. For example, a specific set and arrangement of organs interact in the familiar system of the human body. The system operates in such a way that none of the component organs, or even subgroups of organs, could achieve on their own.

Ecosystems are not designed in any particular way; rather, the way they function is the unplanned outcome of billions of years of evolution. One such function is habitat formation: The combined physical presence of the multiple species in a given ecosystem is what creates the "household" they all inhabit. Ecosystems also filter pollution from water, soil, and the air—not as a service to humans, but because nature is the ultimate recycling system. One species' waste product is another species' lunch. The wider the diversity of needs, the more likely everyone's trash will be useful to somebody. It is only intuitive that more biodiverse ecosystems recycle more effectively—that is to say, waste products such as excrement and dead matter are utilized by other organisms for the nutrients they contain. Imagine a single fish in a bowl: as the fish eats, its waste piles up. Once you add a plant, though, the fish excrement plays a more obvious role in the system. Microbes go to work on the waste, creating food for aquatic plants or algae, which can then be eaten by the fish.

Diversity does not exist only along the food chain from plant to herbivore to carnivore to microbe. There is also stunning diversity at each individual level of the food chain. Evolutionary ecologists are interested in how this diversity came to be and how it is maintained. Why haven't the superior competitive attributes of some herbivore species, for example, enabled them to monopolize plant food resources? Charles Darwin would answer that a single species can dominate only to the point at which other species manage to press back with their own survival strategies.

When multiple species vie for a share of a common pool of resources, they tend to specialize their location, timing, or symbiotic partner, among other variables, to access the resources they need. For example, differing root depths, growing seasons, or tolerance for moisture and shade allow different plant species to stake out resources where and when others are less active. Consider how bats specialize in consuming night-flying insects, while many birds eat insects that are out and about during the daylight.

Such specializations are called niches. From an ecosystem perspective, this broad coverage increases the likelihood that available nutrients will be found and consumed. Indeed, a growing body of research shows that more diverse ecosystems are "more efficient at removing nutrients from soil and water than are ecosystems with fewer species," a finding with important implications for restoring nutrient-polluted bodies of water.[6]

Ecologist Bradley Cardinale wondered whether *the way* more species consume nutrients more efficiently than fewer species is *by way of* occupying different niches.[7] To test this, he created artificial streams to observe nutrient uptake by algae in different conditions. One stream type contained a variety of flow velocities—larger, more easily torn algae could cluster in the slower sections while smaller, more compact algae could dominate the faster currents. He also physically brushed certain patches of the stream to favor fast-growing species that could easily colonize the newly disturbed spots. Cardinale then set up another type of stream with uniform flow, no disturbances, and thus no possibility for niche variation. He varied the number of species per stream from one to eight.

Cardinale found that in the streams with various flow speeds and patchy disturbance, the algae did indeed spread out, each species dominating a unique niche space. In these heterogeneous streams, the most biodiverse algae communities removed nitrate 4.5 times faster than the average rate of removal by a species grown alone (and also faster than the most efficient monoculture). By contrast, "the loss of niche opportunities [in the homogeneous streams] led to reduced diversity."[8] In these niche-poor streams, nutrient uptake efficiency depended on the presence of the species with the most efficient nutrient uptake rather than on algal diversity per se.

It is not surprising that the rich algae communities, which consumed nitrogen faster than their less diverse counterparts, grew more quickly as a result—nitrogen is essential for plant growth. For each additional species in the heterogeneous stream system, the overall biomass (weight of living organisms) increased at a regular interval. Findings that biodiversity makes ecosystems as a whole grow more vigorously have been replicated in numerous studies over the past few decades.[9]

Biodiversity and Productivity

At a time when we are looking to ecosystems to draw carbon out of the atmosphere, the observation that biodiversity boosts ecosystem productivity (the rate of overall biomass growth) is a big deal. Every living being is made out of carbon that was once in the atmosphere and will eventually cycle back through to the atmosphere. Carbon accounts for half of the nonwater weight of trees, which draw carbon out of the atmosphere during the process of photosynthesis. Thus, the more an ecosystem grows and flourishes, the more it draws down carbon to fuel the bustling biosphere.

And it is not only plant diversity that counts for ecosystem productivity and carbon capture. Animals play a key role, too. An intriguing study of a mixed forest-savanna area in Guyana found that mammal diversity is associated with increased carbon concentration in the soil.[10]

The authors of the Guyana study explain: "Large-bodied seed dispersers such as tapirs, peccaries [two pig-like species] and primates ingest, digest and defecate large amounts of fruit pulp and seeds, as well as grasses and leaves, moving plant matter across the landscape and processing it in ways that make it available to a larger diversity of invertebrates, fungi and microbes. Plant and food matter moved by these animals is secondarily buried by dung beetles, by scatterhoarding rodents, or through trampling and rooting by ungulates."[11]

Greater mammal diversity increases overall feeding activity and thus the amount of organic remains on the ground, where it can be converted by microbes into soil organic matter. By increasing soil carbon, the increased mammal diversity may also indirectly increase the aboveground carbon biomass of trees, according to the study.[12]

Another paper, which argues that restoring half of the Earth's surface to intact ecosystems is necessary to avert catastrophic climate change, elaborates on the relationship between biodiversity and carbon sequestration and storage:

> It is no coincidence that some of the most carbon-rich ecosystems on land—natural forests—also harbor high levels of biodiversity. Evolution has generated carbon-rich forests by packing in long-lived trees that also feed stable soil carbon storage pools. This packing effect is made possible by high levels of coexistence among

> diverse species and growth forms, and this coexistence has been made possible by the biotic interactions that generate competition and defense. It is the very pests, pathogens, pollinators, decomposers, and predators that comprise a tropical forest that generated the carbon-rich growth forms (in both wood and soil) that take the carbon out of the atmosphere.[13]

Naturally, the inverse is also true—reduced biodiversity results in lower ecosystem productivity. One analysis showed that an intermediate level of species loss reduced plant production by an amount similar to ultraviolet radiation and climate warming.[14]

Biodiversity and Stability

Another important ecosystem function—one that Miyawaki frequently cited as a reason to plant native species—is the ability of the system as a whole to withstand stressors such as severe weather. Miyawaki's proposal to plant a forest seawall along Japan's northern Pacific coast pivoted not only on native hardwood species' longer taproots, which would help anchor them during a tsunami, but also on the forest aspect: Members of a coevolved biological community interact in ways that are ultimately mutually supportive.

An important aspect of biodiversity is "functional diversity," which refers to the range of traits among different species that drive ecosystem functions. The species that share a functional trait are considered to be members of a functional group. Nitrogen-fixing plants, such as clover, beans, and alder trees, are an example of a functional group. Each of these plants has the ability to form a symbiotic relationship with nitrogen-fixing bacteria that live in tiny lobes called nodules in their roots and turn atmospheric nitrogen into plant-available nitrogen. By virtue of their bacterial symbiont, nitrogen-fixing species make nitrogen available not only to themselves but to the ecosystem as a whole.

"Functional redundancy," or diversity within a functional group, is equally important. Consider a scenario in which there is just one nitrogen-fixing species in an ecosystem and it succumbs to a pest outbreak or protracted heat wave. If this unitary nitrogen-fixing species were to perish, the ecosystem it inhabited would lose the constant nitrogen injection it had furnished.

In this case, functional redundancy would mean the presence of several nitrogen-fixing species in the ecosystem. If one species was wiped out, others could pick up the slack. As explained by internationally renowned biodiversity expert Sandra Diaz: "The larger the number of functionally similar species in a community, the greater the probability that at least some of these species will survive changes in the environment and maintain the properties of the ecosystem."[15]

The importance of biodiversity as a stabilizing force in ecosystems is more important than ever, given the constant assaults on the environment from resource extraction, pollution, and climate breakdown. "There is consensus that at least some minimum number of species is essential for ecosystem functioning under constant conditions and that a larger number of species is probably essential for maintaining the stability of ecosystem processes in changing environments," wrote an international team of biologists in the journal *Science*.[16]

Biodiversity versus Native Biodiversity

In the coastal town in France where I live is a 1.6 ha (4 ac.) "exotic garden" that is elegantly presented as "the Southern Hemisphere in Northern Finistère" (the department, or county, where it is located). The vast botanical collection of 3,500-plus species is drawn from South Africa, Oceania, the Canary Islands, Latin America, and elsewhere. Perched on and around a rocky outcropping over the ocean, the garden is sheltered in a microclimate created by the heat absorbed by the dark granite protrusions, and surrounded by hedges that block cold winds. The Gulf Stream current carried across the Atlantic keeps the coastal area mild; it rarely dips below freezing.

A trail weaves through the site, passing over bridges, under small tree canopies, beside goldfish ponds, up steep metal steps to a lookout point, and around bends overlooking the harbor. The vegetation is multilayered, somewhat dense, even lush; it actually looks rather natural. It is somewhat exhilarating to follow the twisting path, never knowing what charming nook awaits beyond the next corner. The best time to visit is at dusk in December, when the green garden is decorated in lights.

The more I ponder what makes native plants the proper choice for ecosystem restoration, the more my mind drifts to this arboretum, which is the exact inverse of what Miyawaki recommends, and as much a work of fine art as it is an ecosystem. With thousands of plant species packed into a tiny patch of land, it seems to pass the biodiversity test for a healthy ecosystem. Does the botanical diversity, though, translate to faunal and microbial diversity? What sort of food web exists in this environment?

I asked evolutionary biologist Marc-André Selosse, who lives in Paris but knows this garden, whether it could be considered a stable, functioning ecosystem. He clarified that everything is always changing and evolving in nature—nothing is stable in the sense of being static. As for whether the garden is a real ecosystem? "Humans are part of the ecosystem," Selosse answered, "so if it's managed by humans, it's an ecosystem."

Reframing my line of inquiry, he added: "The question is, rather, what would happen if we stopped tending the garden, if we stopped being one of the players of this ecosystem?" Noting that typically only about 10 percent of introduced species are able to survive on their own in a new environment, while 1 percent become invasive, he said that most of the exotic species would die out without human assistance, while a few may escape into the surrounding area. (Curiously, one of the most common plants in town is a towering, conical, purple-flowered, Dr. Seuss–like plant from the Canary Islands.)

If the garden were no longer managed, Selosse continued, "I expect local species would take over most of the introduced ones, with a

Gaurav Gurjar planted the Maruvan mini-forest in 2019 in the desert of Rajasthan, India. *Courtesy of Gaurav Gurjar.*

The Maruvan mini-forest circles two years later, in 2021. *Courtesy of Gaurav Gurjar.*

A newly planted mini-forest lines the perimeter of Iranian dairy farmer Bahram Torkaman's property on the Qazvin Plain, just west of the capital Tehran. *Courtesy of Bahram Torkaman.*

Torkaman's farm mini-forest in 2021, after nearly five years of growth. *Courtesy of Bahram Torkaman.*

The Bulu water catchment area just after planting in March 2020. *Courtesy of Agborkang Godfred.*

The Bulu water catchment area after nineteen months of growth. *Courtesy of Agborkang Godfred.*

Dr. Akira Miyawaki holds up a seedling during Arvin Sango's first planting day in Madison, Indiana. *Courtesy of Mark Gish.*

Shubhendu Sharma on site at Beirut's RiverLESS Forest. *Courtesy of theOtherDada.*

Miyawaki's Yokohama National University forest was at least forty years old when this picture was taken in 2019. *Courtesy of Kazue Fujiwara.*

Cedar trees line the long walkway to the Rinnoji Temple in Sendai, Japan, prior to their removal in 2005 due to the construction of a highway underpass. *Courtesy of Doryu Hioki.*

Doryu Hioki of the Rinnoji Temple decided to replant using the Miyawaki Method. Here, young saplings line the walkway. *Courtesy of Doryu Hioki.*

In clear contrast to the cedars, this fourteen-year-old Miyawaki mini-forest forms a multilayered band of vegetation along the walkway. *Courtesy of Doryu Hioki.*

Employees and community members plant saplings at the Arvin Sango company headquarters in Madison, Indiana. *Courtesy of Mark Gish.*

Keiichi Ohue of Espec Mic in 2019, measuring the growth of forest mounds planted in 2014 at Arvin Sango. *Courtesy of Dan Grady.*

The 2014 and 2016 Arvin Sango mini-forest mounds in May 2021. *Courtesy of Dan Grady.*

Volunteers prepare to plant the second section of Beirut's RiverLESS Forest in November 2019. The first section, along the edges, was planted in May 2019. *Courtesy of theOtherDada.*

A very tiny forest planted in the courtyard of a Beirut school damaged by the 2020 nitrate explosion at the Port of Beirut. *Courtesy of theOtherDada.*

The Beirut Power Plant Forest grows in a traffic median—a previously overlooked scrap of land. *Courtesy of theOtherDada.*

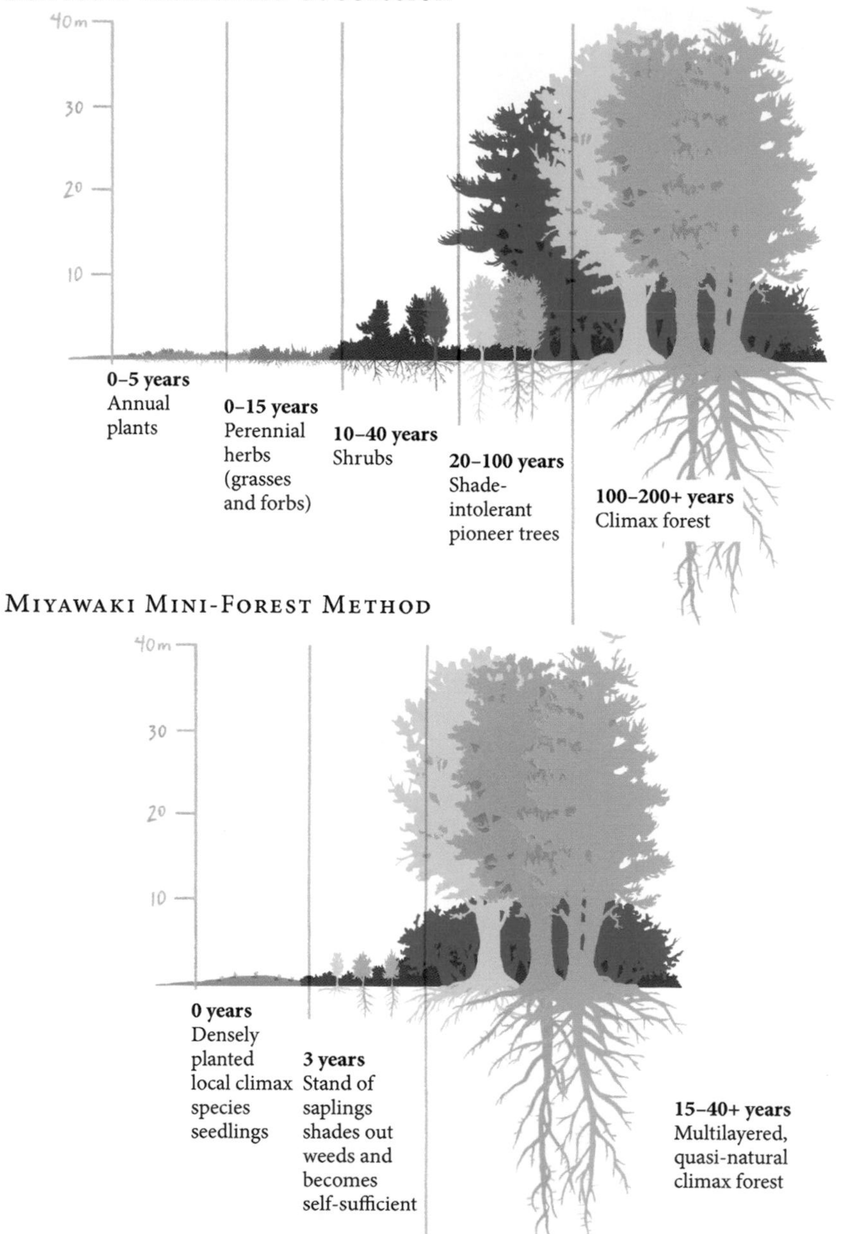

The Miyawaki Method (*bottom*) speeds up the process of natural ecological succession (*top*) through the planting of climax species. *Illustration by Elara Tanguy.*

few exceptions. Also, many of the plants growing there need light, like many plants with nice flowers," Selosse said, adding that many herbaceous plants and shrubs are also insect-pollinated. "You may after some time get to a forest with a few species of trees. And then . . . flowering species are rarer. In this case, you have dense populations where wind pollination is efficient. Really, the plants we cultivate in the garden are not very suited for the climax stage, so I suspect that most species will not be adapted to the late successional stage."

One of the reasons introduced plants die out without human intervention is because local insects do not necessarily pollinate them. "And that's the reason why some species do not produce seeds," Selosse said. "But most of the time there is a bit of pollination even if the insects are not very adapted to these plants, because they pollinate them by mistake—they meander and try. So, it's never zero pollination, but efficiency may be low."

Inversely, some insects die out without the plants to which they are adapted. Many herbivorous insects, such as caterpillars, feed on only a limited number of plant species with which they have coevolved and whose toxins they can tolerate. When these plant species disappear from the landscape, the insects that depend on them disappear too, a loss that in turn hurts the birds that feed their young with those insects.[17]

When I visited the garden one early evening in late July, I saw some insects: flies, gnats, a few bees and bumblebees visiting flowers, and dragonflies hovering around the pond. I also saw several pairs of red admiral butterflies flit by—a nectar- and ripe-fruit-eating species that can be found in gardens, parks, and urban areas throughout Eurasia, North Africa, and North America. In concert with the cooing of pigeons, a handful of European goldfinches sang from the tops of eucalyptus trees. Like the red admiral, this small seed-eating bird has a very wide distribution across Europe, Asia, and North Africa.

Jean-Michel Moullec is the garden's head botanist. He told me that rabbits visit the garden, too, becoming bothersome when they dig up the roots of young plants. Snails and aphids also populate the garden as pests, although the aphids have ladybugs to keep their population at bay, and perhaps the seagulls at the coast eat some snails. A couple of exotic animals that presumably hitchhiked in on exotic plants include a flat worm in the soil, along with the cochineal, a small, flat insect that feeds on cactus and has no natural predators in the area. When a cochineal outbreak occurs, there are not many options other than to zap it with a pesticide. The Asian hornet, invasive throughout France, shows up in late summer and thoughtfully helps with pollination.

A few local birds and bees, too, pollinate plants in the process of drinking nectar. Moullec suspects some bird species have adopted the new plants into their diets little by little since 1995, when the garden was established. Other plants are wind-pollinated, and there is no shortage of wind along the coast.

Making seeds, though, is only part of what it takes to reproduce. The seeds of some species depend on fire to germinate in their native habitats. Garden staff simulate this natural disturbance by subjecting the seeds to a heat treatment. For plants that do not efficiently reproduce sexually, seedlings are produced in greenhouses using cuttings taken from mature plants. Irrigation lines run through the garden, and the giant ferns are misted daily.

Given the level of maintenance needed, even in this incredibly botanically diverse garden, it is clear that biodiversity *per se* is not what makes an ecosystem self-sustaining. Species' interactions with the environment and disturbance regimes such as wildfires play a role. Functional biodiversity matters, too.

"Biodiversity is something that people struggle to understand," Selosse said, "because diversity is not only among species—there is also genetic diversity within species." Furthermore, the vast majority of species on Earth are neither plants nor animals, but microbes. "Often, it's not what we see. We also need our brains to visualize nature. Of course, smells, views, sounds, all of that is giving us a lot of information, and we have to be very open to our senses. But without your brain, you only see the upper crust of the world, and that's the problem with the concept of biodiversity."

Our tendency to focus on visible biodiversity is not necessarily problematic, on the other hand, if it motivates us to protect plants and animals. That's because all of these bigger species are associated with scores of microorganisms, Selosse explained, and if you maintain the big species, you maintain the microbes. By contrast, for example, "when white rhinos are extinct, most of the microbes that are specific to them go extinct as well," he explained.

When a plant is introduced somewhere outside of its natural range, its microbial symbionts may or may not follow, Selosse said, or they may follow many years later. In the meantime, the introduced plant may or may not find local symbionts, such as the mycorrhizal fungi that surround plant roots and help them access water and nutrients from the soil. In the absence of such underground partners, the plant is weakened. On the other hand, an introduced plant may be strengthened by having escaped its native pathogens.

Because of these and many other variables, the changes in biodiversity that result from the "artificial" addition of a new species to a particular local environment cannot not be computed by simple addition. In fact, subtraction is the more relevant mathematical concept at play. Selosse illustrated the likely outcome with a hypothetical example:

> If you introduce 100 seedlings from a wonderful tree from Virginia, for example, and propagate it over Europe, you have to be very naïve to think that you added something to diversity. Because you added one species—okay, but the question is, what did you destroy on the way? Sometimes you may reduce the diversity of another species that may be a local competitor of the introduced species. And if you reduce the local one, even if it doesn't disappear, the reduction in the number of individuals is already a genetic loss.

Furthermore, Selosse continued, the 100 specimens imported from Virginia represent only the narrowest slice of the genetic diversity inherited by that species as a whole.

An exotic garden is unlikely to increase or improve biodiversity over the long term—not even locally—in spite of thousands of added species, although that is not the purpose of an artistic and cultural site such as this. Rather than boosting biodiversity, the unintended consequence of species introductions is that biodiversity is stirred up in a way that creates new winners and losers. Planetwide, 13,168 plant species are known to have become established outside their native ranges, a process that results in a global erosion of genetic, functional, and species diversity, ominously referred to as biotic homogenization.[18]

Large numbers of rare endemic species and locally adapted populations attuned to very specific conditions are replaced by smaller numbers of "cosmopolitan" generalists that can survive a wide range of conditions. In this way, the planet loses not only regional distinctiveness, but also thousands of years of evolutionary fine-tuning. The hasty global reshuffling of species and ecosystems combined with the rapid recalibration of local weather patterns due to climate change is simply an extremely risky experiment.

That native plants are seen as a safe bet for restoring ecosystems is not only because they are adapted to local environmental conditions but also because natives have greater potential to interact with one another, and with local fauna and microorganisms, in ways that make the whole ecosystem more productive and resilient to external stress. Ecosystem restoration is fundamentally about the restoration of ecological interactions, which necessarily implicates native species.

Tree Fervor

Planting trees is a popular "natural climate solution," and rightly so. Trees store carbon and feed the soil with carbon compounds. Excess carbon in the atmosphere is a key ingredient for regrowing Earth's lost ecosystems. Largely due to forest clearing, the world's soils have lost 27 percent of the organic carbon that was present prior to the spread of civilization, according to one estimate.[19]

By another estimate, anthropogenic carbon emissions due to deforestation and related land-use changes from the advent of agriculture (circa 8000 BC) through 2030 is about 486 gigatons, which is nearly equivalent to the estimated total emissions from fossil fuel combustion between 1750 and 2030.[20] The loss and degradation of vegetation and soil represents a massive transfer of carbon from the land "pool" to the atmosphere pool (and also to the ocean pool, where dissolved carbon leads to acidification).

It is possible, however, for the land to regain much of the carbon it has lost when ecosystems regenerate. Dawning awareness of this most elegant of solutions is driving a well-placed tree-planting fervor from wherever you might be to the planet's farthest meridian. Three hundred fifty million trees planted in a day here and 40,000 in a month there—ambitious goals like these make headlines. The beloved, worshipped, long-living, ever-giving tree symbolizes a way to put things right again. Ten thousand years ago, there was an estimated 2.5 billion ha (6.2 billion ac.) more forested land than there is today—an image that hints at what our world could become once again.[21]

Presumably inspired by such a vision, a Switzerland-based team of scientists mapped out the planet in terms of land area that is climatically suitable for forests. They found that forests could grow naturally on 8.7 billion ha (21.5 billion ac.)—about two-thirds of land on Earth—under current climatic conditions, whereas forests currently exist on 5.5 billion ha (13.6 billion ac.).[22] *Forest* is defined by the Food and Agriculture Organization of the United Nations (UN) as any piece of land at least .5 ha (about 1 ac.) in size, covered at least 10 percent by trees. The Swiss team released their findings in 2019: After excluding 1.4 billion ha (3.5 billion ac.) of potential forestland that is currently used for cropland and cities, the remaining 1.8 billion ha (4.4 billion ac.) of largely degraded land has the potential to become forest.[23]

"When our research was accepted to be published in the journal *Science*," co-author Thomas Crowther says in an online TED Talk, "nothing could have prepared us for the media explosion that followed. Suddenly, it seemed like the whole world was talking about the potential of trees."[24]

Published in the lead-up to the UN's Decade on Ecosystem Restoration program, the study was a catalyst for action. The World Economic Forum launched its Trillion Tree Campaign in 2020, while the European Commission pledged to plant three billion additional trees in the European Union by 2030.[25] Alongside the enthusiasm, however, came a stream of criticism. The study's conclusion about ecosystem restoration potential had been transformed by the media storm into a simplistic message about tree planting as the silver-bullet solution to climate change, Crowther explained.

The danger of focusing too much on trees is fourfold: (1) As Miyawaki emphasized, it is not only trees that make up a multilayer natural forest; a managed Douglas fir plantation is not a forest; (2) Planting trees to create new forests—even using the Miyawaki Method—is no substitute for protecting existing, mature forests; (3) Replacing native grasslands with forest is a misplaced effort that damages those ecologically important ecosystems; and (4) Planting trees is no substitute for reducing overconsumption, which not only generates greenhouse gas emissions but also further destroys and degrades ecosystems through endless mining of raw materials and disposal of waste on land and in oceans. "This view of trees as an easy way out is such a tempting perspective, but it is a real threat to the climate change movement and to the ecosystems that still remain," Crowther lamented.[26]

Crowther opened his lecture with a recording of the soothing chatter of birds, frogs, and insects in a natural forest. He then played a recording of a breeze rustling the leaves of a eucalyptus plantation, void of any of the voices of forest fauna. "The UN suggests that almost half of reforested areas around the world are monocultures just like this, planted for rapid timber production or carbon capture," he said, explaining that the eucalyptus plantation is not an ecosystem at all, but rather a tree farm—commercially useful, but counterproductive to ecological restoration goals.[27]

Plantations make up 3 percent of the world's total forests and nearly half of all planted forests; 93 percent of global forests are naturally regenerating, as opposed to having been planted. Primary forest, also referred to as old-growth forest, is defined by the UN as "naturally regenerated forest of native tree species, where there are no clearly visible indications of human activities and the ecological processes are not significantly disturbed."[28] Such forest, located mainly in tropical and boreal regions and including areas inhabited by indigenous communities, makes up about a third of global forestland. This suggests that two-thirds of the world's forests *are* significantly disturbed by anthropogenic activities, such as to logging.

Because much of European forestland is managed for timber, for example, a 10 percent increase in total forest area since 1750 "has failed to result in net CO_2 removal from the atmosphere, because wood extraction released carbon otherwise stored in the biomass, litter, dead wood, and soil carbon pools."[29]

Planting trees to protect the planet is thus an extremely vague proposition, if not woven into an overall strategy to conserve and restore ecosystems. Since 1990, planted forest area has expanded while primary forest has shrunk. While we plant trees, mature forests are cut, and young forests are not allowed to grow old.

Forever Young

The United States is the world's number one producer of industrial roundwood (raw logs, chips, and wood residue), wood pellets and other agglomerates, and pulp for paper.[30] Most of the raw wood comes from the southeast region of the country. Production of wood pellets for energy is a fast-growing industry driven by European demand. Enviva, a major biomass producer in six Southern states, produces more than 5.3 million metric tons per year of pellets, harvested from trees growing within a wide radius of each of their ten facilities.[31]

"Every day, there are lines and lines and lines of trucks," Holly Paar told me. Paar is one of the directors of the Dogwood Alliance, an organization that partners with Southern communities to slow down the timber industry and let more forests grow old. The trucks she referred to are used to haul freshly harvested logs to the Enviva facilities. "It's a constant process, all day and even into the night." The surrounding landscape thus becomes one of loblolly pine patches, each patch containing trees of a single age based on the timing of the previous clear-cut. There is barely anything "that remotely looks like a forest, and of course no virgin forests anywhere," Paar said.

Because loblolly pine and other plantation species grow quickly and can be harvested in twenty to thirty years, and because the timber industry is very active in the Southern United States, Southern forests have a permanently young age structure.[32] Half are less than forty years old, and a little over a quarter are older than sixty years. "Less than 10% of Southern forests are older than your grandmother," goes a slogan coined by Dogwood Alliance.

Between 1630 and 1910, the United States lost about a quarter of its 420 million ha (1.15 billion ac.) of forest, according to the US Forest Service, after which the amount of forested land stabilized.[33] In addition to total acreage, the character of US forests has also changed dramatically over the past couple hundred years. In the US South, 86 percent of forest is timberland, while just 1 percent is legally protected from timber harvest; a slightly higher 10 percent of Southern *wetland* forest is protected from logging.[34]

The fact that US (and European) forests are largely managed for timber means that trees are actually planted very frequently—to make up for what is frequently cut. It is possible that nobody loves to plant trees more than the timber industry does, given that their business model depends on it. Paar explained that because people generally have a romantic attachment to planting trees, it is easy to pass off forest biomass as a form of "green energy." It works as long as public attention is trained on the new trees that are being planted rather than the older ones being cut.

The public is missing the boat, though, Paar said. Because "actually what you're doing when you're not keeping forests in the ground, when you don't let trees stay old—you're playing a catch-up game that you will never win, no matter how many trees you plant." Forests harvested in an endless thirty-year cycle are forever young, and because they have had less time to grow, they have had less time to accumulate carbon in the form of biomass (wood, leaves, and any other living tissue) and fewer chances to interact with other forest-dwelling species.

On account of their small size, younger trees not only store less carbon, they also accumulate carbon more slowly. Trees with trunks that measure about 100 cm (39 in.) in diameter typically add an average of 103 kg (227 lbs.) per year of aboveground dry biomass (not counting water weight). "This is nearly three times the rate [as] trees of the same species at 50 cm [20 in.] in diameter, and is the mass equivalent to adding an entirely new tree of 10–20 cm [4–8 in.] in diameter to the forest each year," write the authors of a 2014 study.[35] In terms of carbon drawdown, it is as if every healthy large tree "plants" a new tree every year.

Old forests with big trees thus have a key role in mitigating climate change through carbon storage and drawdown, while also buffering the local impacts of climate breakdown. Wanting policymakers to grasp the significance of these ecosystem services, Dogwood Alliance calculated the economic value of protecting wetland forest, estimating that it is fifteen-fold greater than the value of harvesting that forest for timber.[36] The high value of intact wetland forests is due not only to income from recreation and tourism but also to the costs they mitigate. Wetland forests provide protection from natural disasters like flooding, and they clean the surface waters that municipalities draw from to feed our taps.

The value of older forests is recognized by some in Sweden, too, where 71 percent of the lichen-rich boreal forest area has been lost in the past sixty years and has been largely replaced by younger stands.[37] Lichen, more abundant in older forests, is an important winter food source for migrating reindeer herds, and thus is critical to the livelihoods of indigenous Sami herders. Sweden is the third largest global exporter of wood pulp, paper, and sawed timber.

Bioenergy accounts for a smaller 10 percent of forest production, but this is likely to expand as a development priority for the Swedish forest industry.[38]

In response to the industrial pressures on Swedish forests, environmental and youth movements and Indigenous groups wrote an open letter to the European Commission in 2021. The letter was a plea for forest biomass burning to be removed from the Renewable Energy Directive II, which aims to increase renewable energy sources to 32 percent of Europe's total by 2030.[39] Calling forest biomass burning a "false climate solution," the authors argue that burning biomass actually releases more CO_2 than does coal burning per energy unit produced. "There is no question—the fossil era is over and we must immediately phase out fossil energy. The goal, however, must be to stop combustion, and not to replace one carbon source with another," they state.[40]

"Natural forests are not renewable. Trees can be planted, but not forests," the authors write, urging that mature forests should be left intact to minimize climate breakdown, and that young plantations are not an adequate substitute. "If you plant wheat, you get a wheat field, not a meadow. If you plant pine trees, you get a timber field, not a forest. Real forests are complex ecosystems, a bedrock of a multitude of life and home for many species."[41]

Our Home on Earth

In his 2020 book *Fanning the Embers of Life*, French philosopher Baptiste Morizot points out that as humans, we tend to see ourselves as being categorically separate from "nature," leading us into either of two opposing camps: those who want to exploit it or those who want to protect it from human exploitation by setting aside nature reserves.[42] This dualist thinking suggests that what is given to nature is taken from humans and vice versa. Morizot reminds us that we are actually part of nature, and that it is possible to take what we need, like any other animal, without overexploiting the environment.

One example of how we obscure our relation with the rest of nature is how we speak in terms of "producing" food through agriculture. However, if to produce something is to design it and to mold a passive material according to that design, then no human has ever "produced" a grain of wheat.

"The prodigiously complex organic form of wheat originates from millions of years of evolution. Early human domesticators simply selected the biggest grains and ones that did not easily break off the ear, two modifications derived from spontaneous variations in wild wheat," Morizot writes.[43] Wheat is "produced" by photosynthesis, he continues, modulated by the long evolution of cereals that resulted in their ability to turn sunlight into living tissue.

The Western world's broad acceptance of the concept of agricultural "production" is not merely a semantic issue. Morizot explains that it is a cultural mythology that relieves us of any sense of debt toward the ecosystem processes that generate the plant and animal species populating our fields and pastures. "The myth of production cancels the necessity for reciprocity with regard to the donor environment," he states.[44] It objectifies other life-forms, putting people separate from and above rather than within the circle of life.

Rather than producing life (although we reproduce our own genetic lines), people have *been produced by* the ancient web of living things interacting in ecosystems over a period of time beginning well before *Homo sapiens* appeared less than a million years ago. Living ecosystems created us and continue to sustain us.

Morizot suggests that living in balance with the rest of the biosphere requires acknowledging our common heritage with every one of our fellow inhabitants. Our extended family arguably encompasses more than eight million species; we need never feel lonely. The more we are able to see ourselves as legitimate members of the ecosystems we inhabit, the more our ties to other species become apparent.

Any dietician knows we need microbes in our guts to help break down food into nutrients. But do we need creatures like beavers, for example?

It might come as a surprise to many that these constructive rodents help us to breathe. We are connected to beavers indirectly through a chain of interactions spanning at least two ecosystems. When beavers build dams in river deltas, they create tide pools where juvenile salmon seek refuge in times of drought and hide from predators before venturing into the vast ocean.[45] Once in the ocean, salmon feed whales, whose nitrogen- and iron-rich excrement feeds phytoplankton, which in turn generate most of the oxygen in the atmosphere (not to mention that algae also sequester carbon and constitute the base of aquatic food webs).

In fact, it is usually not individual species that we need so much as the interactions among a great many species. It is this diversity of beings coexisting in overlapping groups within locally specific environmental conditions that weave Earth's mantle into a living tissue. To elaborate on the cloth metaphor, imagine a single species as a thread. No matter how lovely the color or texture of that thread, it is fragile and rather useless on its own. Only when woven together with multiple threads does it become a functional fabric, which is stronger the denser the weave. On the other hand, the loss of any one thread can begin to unravel the cloth.

Our Role

Fanfare around planting trees is vulnerable to the same simplification as is climate change discourse in general, which verges on boiling the problem of ecological collapse down to a single guilty element: carbon. It is important not to focus too much on carbon or any other individual component of the greater system if it keeps us from seeing the system as a whole. Efforts to mitigate climate change should not become hell-bent on locking up villainous carbon and throwing away the key—whether through high-tech solutions or with trees and forests.

It is not the job of humans to move carbon around or to immobilize it—that important role belongs to Earth's biosphere as a whole. As Miyawaki and Box noted: "The major structural elements of plants and animals are those that make up the atmosphere: oxygen, hydrogen, carbon, and nitrogen."[46] There is constant exchange between life and the heavens. Carbon is meant to cycle between the earth and the atmosphere, and yet widespread ecological degradation and fragmentation have weakened the planet's green pump and shrunk its green reservoir, contributing to the buildup of CO_2 in the atmosphere and ocean acidification. Our role, rather, is to try to understand ecosystems—how they affect local and global climate systems and how they protect and are protected by the wild diversity of species, including humans, that compose them. And based on this understanding, our role is also to honor ecosystems—including their human inhabitants.

That the human species is part of nature and deeply linked to a wide web of species—the web itself sustaining us—points to the relevance of the Miyawaki Method. Any one of us cannot do a lot to repair the Earth, but we can do something. We can team up with our neighbors, get to know the species of our local ecosystems, and re-create places for them to live—even on as little as a 200 m^2 (2,153 $ft.^2$) patch of degraded land.

A space as small as that can become home to more than 500 species and draw down around 250 kg CO_2/year (551 lbs./year), which is equivalent to the amount of CO_2 emitted during a drive from Amsterdam to Barcelona.[47] A space as small as that can serve as a cooling station on a scorching day, or as a sponge on a stormy day, redirecting rain into groundwater storage instead of shuttling it out of the city in a concrete flume to the sea. A space as small as that can take on a special meaning for the people who planted it and those close enough to watch it mature.

And if a single 200 m^2 (2,153 $ft.^2$) Miyawaki mini-forest can do all these things, imagine what dozens throughout one town could do. What hundreds in *every* city could do?

CHAPTER EIGHT

Mini-Forest Field Guide

We choose the main tree species and their companion species from the potential natural vegetation of the area, collect acorns of those species, grow the seedlings in pots until the root system fully develops, and mix and plant them closely together following the system of natural forests.

—Akira Miyawaki[1]

In this chapter, I cover the basic process for tackling a mini-forest project in your own community. Planting a mini-forest using the Miyawaki Method can be accomplished by following five main steps:

1. **Forming a team, securing funding, and finding a site**. Since many mini-forest projects are collaborative endeavors in public spaces, this preliminary step is essential. We will look at a number of ways to approach these tasks.
2. **Making a species list.** Once you have a team, funding, and a site, it is time to go in search of potential natural vegetation. While this step may be the most complex, it is also a fun opportunity to forge a greater connection with the mini-forest you will be

planting. We'll look at how to make this easier by engaging people who bring a knowledge of local ecology to the table.

3. **Preparing the soil.** This step involves determining the quality of the soil at your site and improving it as needed. Healthy soil is a recipe for success.
4. **Planting.** This culmination of months of preparation can be whatever you make it. However you go about it, planting day is an opportunity to engage the entire community in a project that they will go on to cherish.
5. **Maintaining the mini-forest.** While maintenance is critical, it does not require a lot of effort. After three years, a Miyawaki-style mini-forest is self-sustaining and requires no further maintenance, but ongoing community engagement is important to keep the mini-forest relevant to future generations.

This chapter draws on a number of existing guides authored by some of the organizations mentioned throughout the book (see resources, page 167), as well as on the insights and information shared with me by the people featured herein. It is informed by people I know, like my sister-in-law Aleigh Lewis, who was working on a Miyawaki project in her Los Angeles community while I was writing this book, and my colleagues at Biodiversity for a Livable Climate, who planted a mini-forest in Cambridge, Massachusetts, in 2021. A lot of work goes into planning and executing a mini-forest project, and my hope is that this guide breaks it down into manageable pieces. As with anything that requires a bit of elbow grease, the process is highly rewarding.

Forming a Team, Securing Funding, and Finding a Site

These essential steps to get you started can be tackled in a number of ways. Typically, these tasks are so intertwined they need to be worked on at the same time.

Forming a Team

The first step in a mini-forest endeavor is to form a project team, which may grow over time as the needs of the project change. The right partnership mix is often the key to a well-run project. To get started, consider assembling a group composed of individuals who have different areas of expertise and important personal qualities, including motivation, determination, and networking skills. The team should include someone with landscaping expertise and someone who knows the local plant species and their ecology. Project partners may also include local schools, community gardens, neighborhood councils, city government, parks departments, state forest services, botanical societies, tree nurseries, landscaping companies, and Indigenous tribes.

The team can begin the project by looking for opportunities for funding, or it can begin by finding and approving a site. The order of these steps depends on the project itself. It could be that you or a team member already has a site in mind, or that funding has been secured prior to the creation of the team.

Securing Funding

This process of securing funding will be different from project to project. In France, the cities of Paris and Toulouse engage citizens in solving local environmental and social challenges by inviting residents to propose small public projects that the city then funds. The Miyawaki mini-forests in these two cities were launched in response to such calls for proposals.

Other funding opportunities may be less obvious at first, but they can be found by making connections, asking questions, and communicating a clear message about the environmental and public health value of a mini-forest. Your message may resonate with a city or potential funder's existing goals. Take the example of Daan Bleichrodt's mini-forest in Zaandam (see "An Earthy Education," page 57). Bleichrodt told me: "Municipalities struggle to communicate about climate adaptation and biodiversity.

This project allowed them to show [residents] what they mean by abstract concepts like 'heat island effect,' water retention, and carbon sequestration." Similarly, he said, "large corporates are known to invest if they have local branches or construction activities that cause nuisance amongst local communities," which a mini-forest could potentially mitigate. Miyawaki's collaboration with Japanese industrial corporations, such as Nippon Steel, certainly leveraged this kind of opportunity.

Bleichrodt, whose business administration degree seems to have been put to good use in the fundraising aspect of mini-forest creation, organized an event for large donors and other stakeholders to meet Sharma and learn about the Miyawaki Method. "This evening worked really well," he said. "All stakeholders present got involved in the project and helped me fund the beginning of the program, the first pilots, and the scientific research. We also met Hoek Landscaping, who has been indispensable in the process." This initial funding allowed the first few mini-forests to be planted as a proof of concept, after which the program grew quickly in popularity and received further funding.

Finding a Site

While finding a location for a mini-forest depends on land availability and other logistical considerations, identifying a suitable site depends first and foremost on your goals for the project. For Miyawaki, the ideal planting sites were patches in stressed urban and industrial environments, such as the space around shopping centers or factories, along highways, or on steep, erosion-prone slopes. In such contexts, an "environmental protection forest" is intended as a solution to noise, high temperatures, soil erosion, and what Miyawaki described as the "sometimes-stark urban viewscape."[2]

The main goal of IVN Nature Education in the Netherlands, on the other hand, is to reconnect people—especially children—with nature. Their mini-forests are thus located at schools, public parks, community gardens, or other such places where people congregate. Every mini-forest they create is adopted by a school. Contrast this to the mini-forest along Paris's beltway, known as the périphérique. This mini-forest blocks pollution and noise for adjacent neighborhoods in this dense and populous city, and it improves the ecological quality of the beltway's vegetated margin as a wildlife corridor. Moreover, this mini-forest is sited on what amounts to no-man's-land, which cannot be used for anything else anyway.

For public projects, finding a site is a social process, meaning that a successful project involves input from local government and likely from other stakeholders. If the project is on private property, then the main considerations are physical ones.

The Miyawaki Method can be applied on sites with any prior type of land cover, including lawn, gravel, or asphalt. Bear in mind that a more developed plot will require more effort and equipment to prepare the soil for planting. The prospective land should be clear of underground and aboveground infrastructure, such as electrical lines and pipes. The edges of the planned planting area should be at least 1 to 2 m (3 to 6 ft.) away from any buildings or structures to allow space for the trees to branch and root out.

Other logistical considerations include accessibility and storage space. It is likely that your site will need to be worked with a backhoe to decompact the soil, and that a truck will be needed to drop off compost, manure, and straw. These vehicles will need to have space to access the plot. The site should include space next to the planting area to pile up the aforementioned amendments so they are easy to access when it's time to prepare the soil. Depending on the climate where your site is located, you might require a water source such as a tap or a rain barrel arrangement. Or perhaps your city's park service has a mobile water tank that can pass by the young forest to water it as needed during the first couple of years.

The Miyawaki Method works in spaces small and large. Miyawaki wrote that his method can be applied in strips as narrow as one meter wide: "For construction of an urban forest, a large space where many trees can be planted is desirable. But since we define a forest to be a collection of trees with multilayers, a miniature urban forest can be planted in even 1-m-wide strips."[3] However, the smaller the patch or strip, the more it resembles merely the *edge* of a forest. "A true forest must be extensive enough to have an interior, with a microclimate buffered from extreme conditions outside," Miyawaki and Box explained.[4] "In many cases there is not enough space to retrofit a true forest into an already densely built-up urban or industrial area," Miyawaki and Box continued, "so what is constructed is more accurately a 'green screen,' greenbelt, grove, or other smaller woods."[5]

Part of what creates an interior space in a larger mini-forest is the "mantle community," which is "the denser growth on the forest's exposed edges" that blocks wind and sunlight, contributes to the "creation and maintenance of the shadier, more humid microclimate inside the forest," Box wrote.[6] It's helpful to consider a forest as a large building, with exterior walls (the mantle), a ceiling (the canopy), and a few floors (shrub and subcanopy tree layer). In the densest urban areas, however, a whole forest "building" may not fit in the available space. Afforestt and IVN Nature Education recommend a minimum width of 4 m (13 ft.) for something that begins to feel like a forest. In the summertime, a 4 m wide forest strip that is a few years old cannot be seen through from one side to the other—an attribute that Bleichrodt notes enhances the experience of being in nature even in the middle of a city.

Making a Species List

Once you have a team, a site, and an avenue for funding your mini-forest project, it is time to solve the main puzzle: What are the local climax species? As you'll recall, these species represent the final stage of ecological succession—the shade-tolerant, long-living species that you'd end up with if you left an area with average soils undisturbed indefinitely.

The first step may be simply to figure out what tree and shrub species are native to your climate region. There are many sources to turn to for information: Botanical societies or conservatories, universities, and state park or forest services may be good places to look. In the United States, some states' departments of natural resources keep track of native plants and/or plant communities, and they may even supply such species through state-run nurseries.

Historical texts and art specific to an area can also offer clues about local forest species. How far back do you need to look to know that you are identifying "original" vegetation? For practical purposes, native species in the Americas can be defined as those present in an area prior to European colonization; or, in Europe, prior to 1500, when trade between Europe and the Americas began.

Once you have a solid list of native species, the next step is to whittle that list down to your site's potential natural vegetation. Expert assistance from a plant ecologist is highly recommended at this stage—this person will be familiar with the process of mapping local vegetation types and communities. Any given landscape may have a few different climax communities, depending on soil, topography, and microclimate. So identifying the potential natural vegetation means figuring out which climax community is suited to your site.

Start by taking note of the conditions at the site, such as its elevation, soil type, and exposure to wind or salt spray. Typically, a "climatic climax" forest will grow under average conditions within a given climate region—that is, relatively flat land with mesic soil (not too wet and not too dry). In sites with more extreme conditions (poorly drained or shallow soil, for example), "edaphic climax" communities develop, which may contain smaller, shorter-lived species that would be considered pioneers in a climatic climax forest. Again, the theoretical "climax" refers to the maximum vegetation a given area of land can support under local conditions (which, in some places, may not be a forest at all).

IVN Nature Education's mini-forest handbook includes lists of four different forest communities for the Netherlands based on soil moisture and richness in nutrients. An oak-beech forest could be planted in poorer soils or in richer dry soils; elm-ash and ash-alder in moist, nutrient-rich soils; and riparian forests in any kind of wet soil found along rivers. Each of these forest types has a corresponding species list, and several species appear on more than one list because the communities overlap on the landscape.[7]

Knowing the conditions of your planting site will provide a basis of comparison to natural or seminatural areas with similar environmental conditions. You can survey the species in these comparable sites for an idea of what plants would grow spontaneously (naturally) at your site. In addition, secondary sources will be helpful. For example, the Pennsylvania Natural Heritage Program, in conjunction with the Pennsylvania Department of Conservation and Natural Resources, describes the species composition of more than 100 plant communities found throughout the state, as well as "their associated soil types, geology, related plant communities, and range."[8]

Similarly, the University of Southern California created a map of the hypothesized potential natural vegetation of the Los Angeles River watershed, which includes over a dozen vegetation groups found in Southern California's coastal region. Looking at any particular site on this map gives at least a ballpark idea of its PNV (potential natural vegetation).

Thinking Like a Plant Ecology Ninja

In the middle of a sprawling metropolis populated with many introduced plants, or within an expanse of continuous cropland, it is often far from clear what natural climax vegetation the land could support. In these areas, Miyawaki wrote, "Most of the vegetation has changed under the influence of various different human activities, and most of the real forests made up of native vegetation have been lost." Determining the PNV is thus "so difficult," he said, "that I first thought you needed some special ninja-type skills."[9]

What kinds of ninja skills might be in order? A big one is the ability to distinguish between climax and earlier successional or pioneer species—the typically smaller, shorter-lived plants, including some trees, that are among the first to colonize bare ground. Table 8.1 summarizes the basic distinguishing characteristics between the two.

Table 8.1. Pioneer and Climax Species Characteristics

Pioneer species characteristics	Climax species characteristics
Shorter-living	Longer-living
Faster-growing	Slower-growing
Smaller size	Larger size
Abundant, wind-dispersed seeds	Larger, animal-dispersed seeds
Not shade tolerant	Shade tolerant
Grows in poorer soil	Grows in richer soil
Shallower roots	Deeper roots

When Gaurav Gurjar planted Maruvan, his first experimental forest in Rajasthan, India, in 2018, it included forty-four species. He had gathered the seeds of all the big native trees he found in the area. Several grew much faster than others, and they also quickly attracted the attention of some horse-sized nilgai or "blue bull" antelopes, that "barged into the forest" by jumping over the seven- to eight-foot fence surrounding it.

The antelope ate the tops off all the tallest plants, stunting them, while going easy on the shorter plants. "Whatever climax forest species were there, the antelopes just nibbled on them, and these climax species became much stronger and bushier," Gurjar said. Next came the wild boar in search of something tasty, which again turned out to be the faster-growing trees. The pioneer trees were selected by the boar for their "soft wood and tubers in the roots, which are very watery, soft, and good to eat," Gurjar explained. "They dug out two to three feet of root zone. So, after one year, there was only 70 percent survival rate in this forest."

The pioneer plants in this 2018 forest have since mostly all died. Based on what he learned, though, Gurjar pared down the quantity of species to twenty-two slow-growing, shade-tolerant ones, which seem to be the climax group. "In addition to focusing more on climax species, we also started paying more attention to natural guilds and the fringe/mantle communities," Gurjar said. A fringe community is the low-growing vegetation that extends outward into the adjacent open area from the forest's edge, while a guild is a group of two or three plant species that consistently grow close together or are intertwined in the wild (such as a vine and a tree).[10] Gurjar re-created such combinations in the 2020 forest patch he planted. "Compared with previous patches, this patch is coming up denser and stronger, and the plants appear healthier. Their survival percentage is around 90 percent after the first nine months. There is a stark difference in terms of the appearance and structure of the forest," he noted.

Plant Proportions

A final consideration for your species list is the proportion of each plant. Typically, canopy and subcanopy species will make up 70 to 80 percent of the saplings to be planted, shrubs another 10 percent, and the remainder a mix of midsized plants, but these numbers depend on what is observed in the field. Fujiwara, the plant ecologist who studied with Miyawaki, determines not only the identity of species to plant at any new Miyawaki Method project site, but also the proportions of each, by using the *relevé* method to survey remnant forests or woods as close as possible to the planting site. A phytosociological relevé, commonly used in Europe, is a sample quadrat selected within a stand of vegetation representative of a particular plant community. (A quadrat is a small area used in vegetation sampling that delimits a section of a larger stand of vegetation, like woods or forest.) The species within the quadrat are identified and their cover (abundant or sparse, for example) recorded.

The Minnesota Department of Natural Resources (DNR) website states that the relevé method is gaining ground in North America, too. The Minnesota DNR, which itself uses this technique to study vegetation, maintains a database of vegetation plot data from Minnesota going back to the 1960s.[11]

Encouraging other researchers to contribute to this database, the Minnesota DNR created a "Handbook for Collecting Vegetation Plot Data in Minnesota: The Relevé Method, 2nd Edition" to describe how the department's plant ecologists use the method to collect vegetation data. This downloadable protocol may help your team, potentially in collaboration with local researchers, determine the PNV and relative proportions for your planting site (see resources, page 167).

As the existence of the Minnesota DNR database suggests, there may be collections of relevés already completed for your area. For example, perhaps there are relevés for a few types of oak forests, each with its own distinct, recognizable assemblage of plant species: One is found in more exposed sites, and the other near sheltered, calm bays. You could then adopt the list of species from the forest most similar to your own planting site in terms of microclimate and soil.

It is helpful to keep in mind that the key species in the Miyawaki Method equation are the main climax canopy trees, and that the remaining species play more of a supporting role. Fujiwara suggests that if flowering pioneer trees or shrubs are desired for the sake of diversity or beauty, these species could be planted along the forest's perimeter to form a mantle community.

Sourcing Saplings

With a final species list in hand, you are ready to consider how you are going to source those species. Miyawaki recommended collecting seed from local specimens to germinate and grow in small pots that are 10 to 15 cm (4 to 6 in.) in diameter. Alternatively, you can source young saplings from a local nursery that carries native stock grown from locally sourced seed. Explore both public (city- and state-operated) and commercial nurseries.

In France, the "Végétal local" (local Vegetation) labeling program can help in locating an appropriate nursery for a Miyawaki mini-forest project. Launched in 2015 by the national government's biodiversity office, this label verifies the local and wild (as opposed to hybridized or bred) origin of nursery plants on the market. The program's objective is to preserve native biodiversity throughout the country and to ensure the availability of species suitable for ecosystem restoration. By the end of 2021, seventy nurseries and seed suppliers countrywide were certified Végétal local.

If you cannot source all the species on your list, prioritize a shorter list of native species over a longer list that includes nonlocal substitutes, such as non-native members of the same genus (for example, Norway maple versus sugar maple). Above all, prioritize the dominant native canopy species, or the "main" local forest species.

It is important that the saplings be healthy and undamaged with well-developed root systems. Miyawaki strongly encouraged the use of potted saplings for this reason. (In fact, his method has otherwise been referred to as the "potted seedling method.") Saplings can easily be removed from pots without disturbing their roots. You can verify root development by gently pulling the plant out of its pot—the roots should hold the soil in place in the shape of the pot even when shaken.

Many nurseries sell trees with their roots bare (no pot or soil), which potentially exposes the roots to drying and damage. However, bare-root seedlings are typically more affordable, and in Bleichrodt's experience in the Netherlands, they take root and grow well if handled with care.

Preparing the Soil

Before the mini-forest can be planted, you will need to take a few preparatory steps to make the soil hospitable for the young seedlings. Both urban and rural soils often lack topsoil, the top layer of soil that is higher in organic matter. They are also typically compacted. Site preparation thus entails loosening the soil and improving it through the addition of organic material. Miyawaki recommended building up the topsoil to 20 to 30 cm (7 to 12 in.), since this upper layer is where seedlings obtain water and nutrients and is critical for their establishment.

While it sounds rather simple, this step is actually quite technical and typically requires the participation of someone trained to operate a backhoe and someone with knowledge about soil quality. Thus, Bleichrodt recommends consulting a landscaping expert for help during this stage of the process. City parks departments similarly have such expertise and equipment, which could be contributed as in-kind project support. Either way, it is important for project leaders to understand the basics in order to better support the landscaping professionals involved.

Before any soil preparations take place, the planting site should be measured out and demarcated with a material such as ribbon, powder, or spray paint. If your forest has an irregular shape, consider using Google Earth to measure out the dimensions according to the desired surface area.

To avoid any potential soil erosion, compaction, or drying, do not prepare the site much earlier than a few days before planting. Toulouse in Transition, which planted a mini-forest in the southern French city of Toulouse in 2020 and documented the process in a how-to handbook, advises that working the soil with a small backhoe can normally be accomplished at a rate of 150 m^2/day (1,615 ft.^{2}/day). Accordingly, this work could begin a couple of days before planting day on a 300 m^2 (3,229 ft^2) site, and further ahead of time for larger sites.

The ground should be dug and loosened to a depth of 1 m (3 ft.) or less, but generally at least 30 cm (1 ft.), depending on the depth of compaction. In cities, soil is extremely compact by design, Bleichrodt explained, because sidewalks, roads, and other infrastructure must be able to withstand the high pressure of daily traffic. "And soil that has these characteristics does not allow root growth. This is the main problem for urban trees," he said.

In a series of photos, IVN Nature Education's "Tiny Forest Planting Method Handbook"[12] shows the sequence of steps: The backhoe operator excavates the soil and piles it up next to the plot.

Half of the soil is poured back into the trench along with half of the soil amendments, and all this material is evenly mixed together inside the excavated plot. The remaining excavated soil and amendments are then added and mixed. Most of this work is accomplished using the backhoe, although people with rakes can also help spread the material and smooth the surface.

The choice and quantity of amendments depends on the original soil's quality and texture, which is determined by the proportions of clay, sand, and silt making up the mineral component of the soil. Clay particles are the smallest—they bind together tightly and leave very little air space, holding water such that plants cannot easily access it. Clay-rich soil can be hard and dense and cannot absorb water quickly, causing rain to wash off the surface; and when water is finally absorbed, it takes a long time to drain, potentially leading to waterlogging. Sand particles are coarse, causing the soil to absorb water readily and drain just as fast. Silt particles are smaller than sand and bigger than clay, and are thus better able to hold water and make it available to plants.

To determine your site's soil texture, the Afforestt handbook suggests conducting the "ribbon test." The guide instructs: "Get a palm full of soil, make it wet and knead it thoroughly."[13] Then press the soil bit by bit between your index finger and thumb to form a "ribbon." If you can form a ribbon more than 2.5 cm (1 in.) long, then the soil is clayey. If you cannot form any ribbon at all, then the soil is sandy. In between those two extremes, a ribbon up to 2.5 cm means the soil is loamy, which is the easiest type of soil to work with. The soil texture should be checked in the layers beneath the topsoil layer, as well.

Both permeability and water retention are important. To improve the soil's capacity to hold water, sandy soil should be amended with a naturally absorbent material such as compost or a dry local plant fiber (shredded wood, coconut husk, sugarcane, and so forth). To improve permeability and facilitate root penetration, compacted or clay soil should be amended with dry fibrous biomass, such as rice, wheat, or corn husk; straw cut into small pieces; or even imported sandy soil. To improve soil fertility, you can use a variety of finished vegetable composts or dried/aged manures (such as goat, cattle, chicken, or horse). Depending on the condition of the soil, add anywhere from 2 to 6.5 kg/m^2 (4.4 to 14.3 lbs./yd.^{2}) of compost or aged manure.

Mounds and Slopes

Some mini-forest projects are planted on flat surfaces, but Miyawaki and his team found that forests growing in very rainy areas tend to do better on convex (rounded) mounds or berms rather than a level surface. In other words, the whole planting surface is elevated to form a slight hill. This prevents water from pooling up in any small depressions, potentially rotting the roots. In drier areas, little or no elevation may be needed.

If you desire a high mound, you can add non-toxic building debris such as discarded chunks of cement or wood below the ground surface to raise the soil up higher. Simply dig a trench, piling the topsoil on the side of the trench, and lay the debris into it, mixing some of the topsoil in as you go. Layer the remaining soil on top of the mound.

With expert assistance, it is also possible to plant on steep slopes, where the deep tree roots will ultimately stabilize the soil. On slopes steeper than 1:1.8, the ground can be terraced and secured with light retaining fences (made of branches or thin logs) to hold up the vertical side of each terrace step. These steps will also help volunteers access the area during the planting stage.[14]

Removing Asphalt

Removing asphalt or concrete to plant a mini-forest is possible, albeit with added costs. IVN Nature Education's Music Square tiny forest, which replaced a parking lot, cost 50 percent more than the group's standard projects due to the expense of removing infrastructure like lampposts and relaying pipes and cables, plus a half day of labor for removing the pavement. The layer underneath the asphalt was heavily compacted sand, which city excavators removed and replaced with soil from elsewhere. Removing the sand was actually an unnecessary step, Bleichrodt told me, and in subsequent projects requiring asphalt removal, IVN Nature Education has left the underlying sand in place, using it as the basis of the soil. The compact sandy ground is simply loosened and mixed with the organization's standard rations of organic soil amendment. Bleichrodt prefers this less resource-intensive approach, which he considers more sustainable than transferring topsoil onsite from elsewhere.

Planting

Planting day is the culmination of weeks, months, or even years of preparation, and it's worth celebrating. That's probably why Miyawaki referred to these events as festivals and encouraged broad participation by many groups. To accommodate large, multigenerational crowds, organizers need to be well organized.

The best times for planting in temperate climates are early spring and late fall—before or after the ground freezes—to allow the plants to take root before the summer growing period. In climates with seasonal precipitation, planting is best in the rainy season so the young plants can benefit from plenty of water as they get established. The planting day should occur immediately after the saplings are delivered or picked up from the nursery, so that the fragile young plants (especially those with bare roots) need not wait before being planted.

IVN Nature Education estimates that planting 600 trees, followed by mulching, is a full day's work for six to eight adults plus a classroom of schoolchildren. At the other end of the spectrum, 400 volunteers at the Arvin Sango tree festival in Madison, Indiana, planted 4,000 trees in less than two hours. The bigger the crowd, the higher the level of organization required.

All the supplies should be readied and available when volunteers arrive. This includes enough hand trowels to go around, a few digging forks or shovels, the saplings, buckets for dunking the saplings in water, wheelbarrows, and straw. The tools could potentially be borrowed for the day from volunteers, a school, a municipality, or a community garden. Toulouse in Transition suggests assigning one person to assemble the borrowed tools, keep track of them, and return them to the owners afterward.

To ensure a somewhat even distribution of species across the planting area, the plot can be divided into sections with a representative mixture of species allocated to each section (a quarter of the plants from each species, for example, distributed to each of four sections). Alternatively, the plants could be planted in a completely random fashion. To minimize traffic (and trampling) around the plot, volunteers could be assigned to particular sections of the site, and the tools could be distributed proportionally, as well.

Be sure the site is easy to find by posting signs, if necessary, and wearing yellow work vests. In case of rain, some sort of shelter should be available. Consider providing snacks and drinks, especially if volunteers will be on site all day or for several hours. Once volunteers are assembled, welcome and orient them with an introduction to the project and the Miyawaki Method. This brief presentation could include an introduction to some of the tree species that will be planted. You can channel Miyawaki himself by starting your planting festival with his classic call and response "naming ceremony" (see chapter 5).

Next, instruct the volunteers on the planting process. If the plants have been supplied with bare roots, as explained in the Toulouse in Transition handbook, then the roots should first be dunked into a muddy mixture of soil, well-aged manure, and water. This process, called *pralinage* in French, will help the young roots take hold. For potted plants, the potted part should be dunked in water for ten seconds, or until the soil is drenched and no more bubbles rise to the surface of the bucket.

Volunteers will then dig holes one and a half times as wide and as deep as the root ball. While holding the plant within the hole space, volunteers will gently refill the soil around and under the roots, taking care to keep the entire stem, including the base of the stem, above ground; the plant's soil surface should be level with the surrounding soil surface. Next, planters should pat soil around the plant very lightly with their hands (not their feet), to avoid compaction. Plants should be spaced 60 cm (2 ft.) apart (the length of two hand trowels) when planting at a density of three trees per square meter (or per square yard), and volunteers should avoid planting the same species side by side.

Once all the plants are in the ground, mulch is spread evenly over the surface at a rate of 2 to 4 kg/m^2 (4.4 to 8.8 lbs./yd.^{2}). To translate that ratio into practical terms, if your site is 200 m^2 (2,153 ft.^{2}), then you will need at least 400 kg (882 lbs.) of mulch. Straw is the typical mulch material, but dried leaves, corn or barley stalks, and woodchips are potential alternatives. If you choose to work with woodchips, Bleichrodt recommends they be partially degraded rather than fresh, as fresh woodchips may reduce the soil's plant-available nitrogen. To minimize trampling, a relay system can be set up whereby straw is passed from people remaining outside the plot to others scattered within the plot, who then mulch around the area where they are standing. At Arvin Sango in Indiana, volunteers flagged the shortest plants in order to avoid accidentally covering them with mulch or stepping on them.

Depending on how windy your site is, it might be necessary to stitch down the mulch. To do so, insert pegs every few feet along the edge of the plot. A ball of biodegradable twine can then be passed back and forth over the planting mound and fastened to the pegs. Watering the mulch also weighs it down and keeps it in place.

In some places, it may be a good idea to install a simple fence around the plot to keep people and animals (especially deer and rabbits) out while the young and tender plants are developing over the first three years. Once the trees are big enough, the fence should be removed to allow for the free movement of wildlife.

Maintaining the Mini-Forest

Actively maintaining the mini-forest is very important only during the first three years after planting, as the tree roots get established and as the growing canopy begins to shade out weeds. After this initial period, the forest becomes self-sufficient. Maintenance during this initial period includes weeding, watering, picking up litter, and making repairs as needed. It is important to define who is responsible for what in advance. Perhaps the city will water, and a team of neighborhood volunteers will be responsible for weeding on an as-needed basis. IVN Nature Education establishes a particular grade level at the local school to act as the park "rangers," seeing that litter is picked up and also giving informal tours of the forest to parents and community members.

Miyawaki advised leaving the pulled-out weeds to dry on top of the mulch, thereby adding to it. As pioneer plants, weeds grow fast—faster than baby oaks or other climax forest species—and can in some cases shade out the forest plants if not tended to in the early stages. When theOtherForest in Beirut planted 300 m^2 (3,229 ft.^{2}) of native trees in November of 2019, they had no idea COVID-19 was about to strike, confining them to their homes and leaving the baby forest to fend for itself for several months. When a maintenance team finally returned, the weeds had become so tall and thick it took a whole month to pull all of them out. Compared to the adjacent section of mini-forest that had been consistently weeded from the start, Adib Dada said, the November 2019 section looks somewhat stunted, even a couple of years later.

The rate and height of weed growth may also depend on climate, however. In Zaandam, Netherlands, the fact that 595 species that graced the tiny forest in its first few years was largely due to wildflowers that were not weeded out and to the pollinators these wildflowers attracted. "Just remove the weeds that overgrow your saplings, and let wild flowers bloom," Bleichrodt advises.

Watering

The season of the planting makes a big difference in terms of plants' access to water, as Limbi Blessing Tata discovered in Buea, Cameroon. A forest patch that her team planted just before the rainy season grew much better than the one planted at the start of the dry season, even though the latter was hand-watered from time to time.

Outside of rainy periods, the amount of watering required during the first few years depends on the local climate. Based on their experiences in India, the Afforestt handbook recommends watering at least 5 L/m^2 (1.3 gal./yd.^{2}) on planting day after mulching, and frequently thereafter unless the ground starts to feel too wet. The handbook advises watering with a shower fitting at the end of the hose to ensure a gentle application. In semiarid Iran, Bahram Torkaman watered his mini-forest during the first few years at a rate of only 1.5 L/m^2 (.4 gal./yd.^{2}), and only on hot or dry days.

In Paris, the Boomforest mini-forests were watered only a couple of times per year, in the heat of summer, during the first three years. The Toulouse in Transition handbook suggests watering each plant at least 3 L (.8 gal.) directly after the planting, and then once every two to three weeks during periods of drought. In the Netherlands, watering occurs once per week maximum, and only in dry weather. But rather than watering on any particular schedule, Bleichrodt suggests checking whether the soil is moist or dry by feeling under the mulch layer, and watering only if it feels dry there.

The planted forests should never be thinned, pruned, or treated with pesticides, unless branches are protruding into a roadway, in which case they could be sparingly pruned. The developing forests will not need any fertilizer, assuming the soil was sufficiently amended at planting time. Consider putting up a signboard to alert future land managers of the ecological value of the forest, its self-sufficiency, and its lack of need for any type of maintenance.

Miyawaki explains that over the long run:

> Tree mortality should also be expected, and over several decades to centuries, individual trees will die. Some die sooner through natural forces. Such dead trees and withered branches in a forest should be left on site, for they become decomposed, help increase

biodiversity, and promote forest reproduction. In cases where dead trees in urban forests interfere with the aesthetics of particular landscapes, they should be buried in the earth for decomposition or used as railings along paths. They should not be burned. Finally, to provide a more naturalized structure and function to the forest boundaries or along pathways through the forest, flowering shrubs can be planted as mantle communities that keep fallen leaves inside the forest. This will save on maintenance costs, allow nutrients to remain and be cycled in the forest, stimulate soil decomposer communities, and give residents the pleasure of seeing flowers.[15]

Conclusion

The Miyawaki Method is at once a serious ecosystem restoration technique and an invitation to learn about nature. For those of us who accept the challenge of creating a mini-forest, we immerse ourselves in a scientific field experience concerning species identification, soil biology, forest structure and composition, and community ecology. This may be just what the doctor ordered for our highly urbanized species to develop a sense of connection with the multitude of other species upon which we unknowingly depend.

Ecological restoration and restoration ecology are professional and scientific fields that often require a graduate degree. Indeed, Miyawaki's PhD in plant ecology and decades of field experience surely helped him attain a ninja-like ability to identify potential natural vegetation. When properly implemented, the method has proven successful time and again; when not properly understood or followed, a poor outcome risks tarnishing the method's image. Rather than deter, though, the complexity of the Miyawaki Method should encourage us to become curious, to seek to understand the science involved and recruit professional assistance where necessary, and to follow the steps as faithfully as possible. At the end of the process, you may find that you've made new friends, not only with your project teammates but also with a handful of trees whose names you now know.

CONCLUSION

Eco-Restoration Starts at Home

What gives energy to the forest—to a community of trees—is the community of humans infusing it with their own energy.

—BOOMFOREST[1]

In the summer of 2019, soon after discovering the Miyawaki Method via MiniBigForest, I contacted the Nantes-based organization to ask for their suggestions on planting a mini-forest in Roscoff. They sent me some materials that I used to prepare a proposal for my local decision-makers. I started with my children's school, requesting a meeting with the principal to discuss the idea of a mini-forest beside the playground in the school's spacious courtyard. She responded right away saying she liked the idea and that we should meet, but that we would ultimately need permission from the mayor's office.

The mayor's liaison to schools listened to the proposal with some apparent interest and promised that the mayor and his team would discuss it at their meeting the following week. Things seemed to be humming right along! I paid the liaison another visit to see what the leadership team had thought about the proposal, expecting to hear something like: “It’s a nice idea, but good luck finding the money.” I even held out some hope for a small contribution.

I was mistaken. “It was not well received by the committee,” she informed me. She explained that their primary concern was that the trees’ roots might destroy surrounding infrastructure. This struck me as odd—of course, we would avoid land with underground pipes or overhead electrical wires for a tree-planting project.

It is true, though, that much of the land is spoken for in this densely built town. Residential lots are quickly filling in around the ancient stone infrastructure of the city center. Open space exists primarily in the form of vegetable fields and the many large parking lots needed to accommodate summer tourism.

However, I could think of a few grassy areas—one facing the harbor at the edge of the downtown area, and one beside the municipal sports complex. I asked if the concern about destructive roots applied even to open land such as these sites. She shrugged and noted that there was really no interest in such a project on any public land in the city. This was not going to be straightforward.

The MiniBigForest team helped me and a group of interested parents talk through the roadblocks and how to overcome them. We discussed the various people and groups in town who would need to be brought on board for the project to move forward. Communicating the idea to as many people as possible over the coming year would be planting the seeds of more lasting social change anyway, one of the parents noted.

The questions that a mini-forest proposal raises are useful in generating a broader conversation about the climate crisis. Indeed, over lunch with the MiniBigForest team, topics included the specter of future heat waves, the eventuality of France’s famed Mediterranean vineyards migrating north to Brittany, and the burning Amazon rainforest. Yet in this case, it was the discussion of a solution that led to contemplation of the problem—a discussion that empowers and equips rather than paralyzes.

Soon after the lunch with MiniBigForest, I was invited to make a short presentation on the mini-forest project to the school's parent-teacher association. In my halting, heavily accented French, I hustled through my slides showing Parisian students striking for the climate and pictures of Miyawaki-style forests being planted in schoolyards and other small spaces in Belgium, India, and elsewhere in France. I outlined the benefits of biodiverse ecosystems for local cooling, water and air purification, and carbon sequestration.

Following the meeting, the only comment anyone made to me was that I was brave. I felt embarrassed by this comment and the lack of any others, and concluded that my presentation had been a flop. Little did I know, I was soon to have a new ally in Aurore Clément, a fellow parent at the public school who had taken the reins as president of a local environmental organization called Ekorrigans. This newly launched organization was named with a nod to the Brittany region's Celtic heritage, whose folklore is replete with leprechaun-like characters called korrigans. I was invited to join the team and to situate the mini-forest project as one of the group's multiple objectives.

It seemed that we were off and running again.

In Search of a Site

In March 2020, just as COVID-19 was taking over, nationwide municipal elections swept a change of leadership into Roscoff. In clear contrast to the previous mayor, the new mayor was reputed to have a few ecological fibers woven into her worldview. Actually, I can attest to that, having sat next to the would-be mayor at my first Ekorrigans meeting in December. As the group's founding secretary, she called the vote that confirmed me to the board. Call it small-town serendipity. Shortly thereafter, she resigned from our fledgling organization to run for mayor and won.

After another round of communications, this time with the new liaison, we were assured that we would soon be invited to present the mini-forest proposal to the town council. We jumped into preparations. What were the tree and shrub species native to western Brittany, and were there any local nurseries that could supply us? It turned out that the National Botanical Conservatory of Brest (CBNB) had in 2015 published an inventory of all the plants known to grow in the region, including notations about which were indigenous to any of four departments.

All plants were listed by their Latin names, so matching Latin names to common names was a secondary step. Eventually, I identified thirty-four native tree and shrub species, which I sorted by height.

Soon, we got the long-awaited invitation to formally present to the mayor and her core team. After a brief presentation, the discussion began. The elected officials and city landscaper wanted to know about maintenance requirements, how tall the forest would grow, and where in Roscoff a mini-forest could potentially be sited. Someone suggested a grassy patch currently used for parking near the town's iconic St. Barbe Chapel, which is perched atop a rocky cliff over the sea. It was decided that a small group would visit the site the following week.

On a sunny and still Thursday morning in November 2020, I met Aurore, her husband, the city landscaper, and the head of urban planning at the chapel. We surveyed the site, together imagining how a forest might look here, and how it could be situated. We also dug up a bit of soil to examine its texture and quality.

The ground was compacted, sandy, rocky, fairly low in organic matter judging by its appearance, and covered in a thin layer of mowed grass. At the uphill entrance to the lot, a small gully had been carved out by erosion. The parcel was surrounded by a trimmed hedge of feathery tamarix that partially blocked a view of the sea. We envisioned a bulbous triangular forest filling in the north corner of the land, with a path skirting the northern edge of the mini-forest along the hedge, and opening into a clearing. Here, an outdoor classroom with benches arranged in a semi-circle could accommodate school groups.

Anxious to get the project started, I suggested a planting date be set for the current winter planting season. After all, we had until mid-March to properly plant trees. The head of urban planning turned squarely to me, saying he, too, was anxious to get this forest in the ground, but that we need not rush the process—he suggested we plant in another year. I was frustrated at what seemed like a lackadaisical response to an urgent problem. In one year, atmospheric CO_2 concentration would have crept up by at least two ppm, and hundreds more species could have become extinct. For me, the mini-forest was not meant simply as a nice amenity, it was first aid for the planet, or at least for this tiny patch of planet.

Eventually, I realized the urban planning manager was right about waiting a year to plant. When I later called the nursery owner to see if he could supply the 800 plants we would need for our project, he told me he needs to receive big orders in July for plantings to occur the following winter. This particular nursery—Graine de Bocage—would play a big role in our project, as it appeared to be the only one in the region specializing in native plants. So, if he couldn't furnish us this winter, then it was likely nobody could.

Most nurseries in the area proudly supply beautiful exotics, enticing people to spice up their gardens. But Graine de Bocage has the explicit mission to repopulate the Breton countryside with native trees and shrubs. In fact, as the company's name suggests, the seeds (*graines*) they use to grow their plants are hand-collected from the local, semi-wild woody hedgerows that enclose crop fields (*bocage*). Accordingly, this nursery carried most of the species on our list.

In Search of Species

My original list of thirty-four species still needed some refining now that we had identified the specific location for the forest. At this point it was still a somewhat random list of local species, not necessarily a group that would make up a particular forest community. The species on our list spanned a native range that included windy exposed coastline, sheltered estuary bays, flatland, hills, and riparian zones. Thus, while all thirty-four were native to the western part of Brittany, not all of them would be suited to the hyper-local conditions of our planting site.

Urged onward by Aurore, I began the process of verifying our species list by getting in touch with an agent from the National Office of Forests (ONF, or Office national des forêts), Fabien Acquitter. Over the phone, we talked through each species. Based on his feedback, I could cross out half a dozen species, such as the beech tree, which doesn't grow on the coast, and a couple others that grow only inland or to the south. Acquitter also warned me against planting elm and ash trees, which are subject to fungal diseases that are widespread in this region. He was skeptical about the sessile oak (*Quercus petraea*), which is not commonly seen in this part of Brittany.

However, losing beech, ash, elm, and sessile oak left us with only one species of tree—the pedunculate oak (*Quercus robur*)—for the canopy, an important layer making up at least a third of the forest population. Was it normal for a forest canopy to be composed of just one species?

Stumped, I decided to ask Boomforest what to do—with respect both to the forest canopy composition and to several other questions that I was dying to ask. Driven by the belief that communities everywhere should be empowered with the know-how to plant ecologically functional urban forests, they have generously shared their knowledge and experience with me throughout the rather complex project planning process.

"Have no fear!" they encouraged me when I expressed reservations about my ability to design a functional forest ecosystem. According to Miyawaki, the most important qualities for a project leader are determination, persistence, and competence. Leaders do not necessarily have to be botanists, wrote the botanist, although they should call on professionals "when necessary, to carry a project through to its end according to the natural ecological script for creating an environmental protection forest."[2] I called on several professionals for advice, and yet I was taking it upon myself to properly integrate this counsel into a coherent, ecologically sound plan. This strategy left me with a lingering shadow of doubt, which I sought to overpower with the light of continual research and learning.

Speaking with one voice through the organization's email account, Boomforest's co-leaders Enrico Fusto and Damien Saracini responded thoughtfully to many detailed questions I fired in their direction. How to survey surrounding landscape for native forest species in an environment that had been cleared and maintained for agriculture over centuries? How to identify the dominant canopy species of a native broad-leaved forest when the only nearby woods was largely populated by planted exotic conifers? How to water a mini-forest during its first couple of summers without a nearby tap? If we plant a mini-forest on an extremely windy site, will it thrive?

Fusto and Saracini informed me that Miyawaki had planted up to nine plants per square meter (square yard) in the windiest coastal areas of Japan. Thus, we might consider a more conservative increase to four or five plants per square meter on our windy coastal site. The Boomforest team also reviewed my evolving list of species, offering their feedback as Acquitter, the ONF agent, had given me on different plant types.

Unlike Acquitter, however, who had discouraged the inclusion of ash given its potential for fungal disease, Fusto and Saracini urged me to give at least a few of this beautiful species a chance to find their place in this mini-forest community. "The risk is there," they noted, "but the species-rich environment into which they will be planted could certainly boost the ash's capacity to resist disease."

As to the complicated question about canopy species, Boomforest suggested I go ahead and include the sessile oak, albeit in a reduced proportion, even if it is uncommon in coastal Brittany. They also urged me to get back in touch with the National

Botanical Conservatory of Brest (CBNB) to get a scientific opinion on the canopy question in particular and our mini-forest species list overall.

Agnès Lieurade, in charge of plant and habitat studies at CBNB, helped me to further refine my list of species by pointing out that I had combined species that grow in different ecological contexts. For example, alder buckthorn and the 'Le Gall' variety of gorse (a shrub) develop in poor, acid soils typical of heath ecosystems, a kind of shrubland common in Brittany. By contrast, she explained, ash grows in moist, rich soils, and is rarely associated with the aforementioned species. She also highlighted a species on my list found only in the south of the department, whereas Roscoff is in the north, and that had not been observed at all since the nineteenth century. I crossed that one off my list, along with the alder buckthorn and 'Le Gall' gorse, since our site was not a heath environment. Knowing that ash is adaptable to diverse conditions, I left it on the list but in smaller quantities.

As I further adjusted my list according to Lieurade's response, I felt increasingly confident that it was starting to look like the inventory of a real forest community that could once have existed in Roscoff. After having studied the exact location of our proposed mini-forest, Lieurade offered a final injection of solid scientific guidance by referring me to a study of the "phytosociological characterization of coastal oak forests of Finistère," by Frédéric Bioret and Sébastien Gallet.[3]

This short analysis, sponsored by the government as part of the 2010 French Forest Review, describes two oak-dominated forest communities that had been observed in distinct parts of the department. Interpreting this study with reference to the Miyawaki Method and the particular site for the Roscoff forest, Lieurade wrote: "The parcel that concerns your project is situated in a rocky cliff context. . . . The 'natural potential vegetation' is very probably a forest of the type 'pedunculate oak of cliffs.' This type of forest is found on ledges and favors warm and sunny exposure."[4]

Lieurade directed my attention to the study's list of species associated with this particular forest type. Twelve of the species corresponded to the twenty-three still left on my list. Sure enough, sessile oak was not among them, leaving the pedunculate oak the undisputed champion of the forest crown. Lieurade advised me, though, that it was not a problem to have just one main canopy species, adding that my list of species may have been too long anyway. "There are not that many species in a natural forest in our region," she wrote.

I relayed these discoveries back to Fusto and Saracini, who were about as enthusiastic to learn of the oak forest study as I was. In fact, one of the study's authors was known to the Boomforest team by way of a Japanese scientist who visited the Paris mini-forest in 2019 and mentioned Bioret as an expert. Yet Fusto and Saracini stood grounded in Miyawaki's long view of natural history, which accounts for centuries of human interference, noting that the existing vegetation of the region "may be slightly different than the original vegetation since it has been modified by humans (the study even explicitly mentions historic management of the woods in question)." Firewood was harvested from these woods until the mid-twentieth century, according to the study. In other words, the remnant oak forest that was to be the reference ecosystem for our forest planting did not necessarily represent exactly what would have grown in the absence of human interference.

In the end, I decided to keep all twenty-three native species on the Roscoff forest list (see table 9.1), but to favor the twelve that appeared in the Bioret and Gallet list by planting much higher numbers of these species. I sent this final list to Aurore in mid-December 2020, along with quantities and prices according to Graine de Bocage's catalog. She was preparing a budget to begin fundraising. By January, she had already raised nearly €2,000, which would be noted in a dossier for the city council. The council planned to formally vote on the project in February 2021. Approval was expected, given the favorable reception to our proposal, but just in case there was any doubt, we reasoned that two grand already in the hopper should push it across the finish line.

Table 9.1. Roscoff Species List

Canopy and subcanopy (59%)	*Quercus robur*	Pedunculate oak
	Fraxinus excelsior	European ash
	Quercus petraea	Sessile oak
	Prunus avium	Wild cherry
	Ulmus minor	Elm
	Sorbus aucuparia, subsp. *Aucuparia*	Rowan
	Taxus baccata	Yew
	Castanea sativa	Chestnut
	Acer campestre	Field maple
	Sorbus torminalis	Wild service tree
Subtree (29%)	*Ilex aquifolium*	Holly
	Crataegus monogyna	Common hawthorn
	Corylus avellana	Common hazel
	Pyrus pyraster	European wild pear
	Malus sylvestris	European crab apple
	Euonymus europaeus	Common spindle
	Sambucus nigra	Elder
	Mespilus germanica	Common medlar
Shrub (12%)	*Prunus spinosa*	Blackthorn
	Ruscus aculeatus	Butcher's broom
	Ligustrum vulgare	Wild privet
	Ulex europaeus	Gorse
	Cytisus scoparius	Common broom

Two Universes Converge

Happily, the city council did vote in favor of the mini-forest project. It was time now to expand the circle of participation in and ownership of the project. Aurore made a few key phone calls—to the local press, to the principals of the town's two elementary schools, and to an agricultural high school in a nearby town. Shortly thereafter, a journalist came to interview us and take our picture, making mention in her short, well-written article that we were in search of compost and straw.

All the schools Aurore contacted were enthusiastic, especially a teacher from the agricultural high school. Seeing the project as a good way to integrate hands-on experience into her curriculum, she proposed that her students lead the local elementary school kids on planting day, teaching them to plant the trees as one of five concurrent nature-related workshops that the kids would cycle through in groups. She also suggested involving her students on a longer-term basis in monitoring the growing forest for changes in biodiversity and soil quality.

A couple of weeks later, the official agreement on the project was ceremoniously signed at city hall by the mayor and Aurore, and the event was captured on camera by the press. Véronique Croguennec, the environmental liaison who had been our contact throughout the proposal process, made a few prepared remarks about how trees regulate the climate, protect the soil, and generally serve as a common denominator of all life on the planet.

The conversation that followed the formalities was the honest climate-action conversation I had been seeking. Attendees chatted about the ecological importance of trees and forests, and how this project was one small endeavor the community could embark on in the face of climate breakdown. Noting how planting a mini-forest in Roscoff was at once practical and symbolic, Aurore announced that Ekorrigans aimed to encourage such projects in the surrounding area. My parallel universes were beginning to merge ever so slightly.

Just Keep Walking

"When we first started creating 'village forests,' that is, woods in harmony with the local ecological conditions, our voice was like a faint cry in the wilderness," Miyawaki and Box wrote.[5] My experience was similar, but soon enough I found I had been heard. Get started walking and you will soon have company. You may not have many fellow travelers at first, but the momentum grows.

By the time we planted the Roscoff mini-forest in December 2021, we had also gotten word from the mayor's office of a neighboring town that they had decided to plant a mini-forest, too, and wanted Ekorrigans' help to plan it. In fact, stories of communities throughout the region that were also planting mini-forests started popping up regularly in the local newspaper around the time we planted our mini-forest.

In Roscoff, our core team was just two people. In Aurore, I found someone who shared my passion for the environment and whose energy and skills complemented mine, making our partnership so functional that we inevitably became pals. Involvement in the project expanded in circles around us, beginning with the city leadership team, who gracefully picked up the baton we handed them—from listening to our proposal to finding and making a patch of land available to committing city equipment and labor to the task of preparing the ground for planting. By printing a story in its bimonthly newsletter several months prior to the scheduled planting event, the city also legitimized the mini-forest project and gave local people the chance to become familiar with it ahead of time.

When the nursery reported they would not be able to supply one key understory species on our list—the butcher's broom shrub (a petite evergreen with small pointy leaves, red berries, and an unfortunate moniker)—the Roscoff city gardener offered to grow it himself using cuttings from local wild plants. Fellow Ekorrigans members provided moral support and helped locate compost and straw for the mini-forest. One member in particular—Eric Barbou, an avid birder—helped by championing the protection of a coastal wetland area on the other end of town from the mini-forest site.

During the COVID-19 pandemic confinement of spring 2020, he had spotted a pair of western marsh harriers, somewhat rare birds of prey, nesting in the reeds beside a crop field. One morning while the raptors were out hunting, Eric took two members of the mayor's team to visit the marsh so they could see the birds in flight; he wondered what it would take to protect the site where the couple had nested for at least two consecutive springs. Western marsh harrier is listed by the French Committee of the International Union for the Conservation of Nature (IUCN) as a species that "could be threatened if specific conservation measures are not taken."[6]

Although Eric was unsuccessful in securing protection for the wetland due to property ownership barriers, the story of the western marsh harrier and the endeavor to protect their nesting site is part of the Roscoff mini-forest story. The mutually reinforcing initiatives were united by a bigger vision of local ecosystem restoration and protection. Having shared his discovery of the nesting birds with others in town, Eric expanded our awareness of the wildlife that live here too, and of their habitat needs. While these raptors need undisturbed patches of reeds, other birds need tall trees in which to nest and forage—wetland and forest alike are integral components of this landscape.

The Roscoff mini-forest story is also related to the 2021 report by the Intergovernmental Panel on Climate Change (IPCC) that climate chaos is affecting every region of the globe. Six years prior, world leaders convened in Paris and agreed that, while dangerous, 1.5°C (2.7°F) of global warming was a tolerable limit. Yet, in the absence of rapid and deep emissions cuts, this threshold is likely to be breached by 2040, before my twins' thirtieth birthday.[7] Warmer temperatures transfer more water into the atmosphere, drying land and filling the skies with liquid ammunition for megastorms and flooding.

I think the mini-forests growing in Iwanuma, Beijing, Madison, Roscoff, Buea, Paris, Barking and Dagenham, Zaandam, Beirut, on the Qazvin Plain, along the Luni River, in rural Maharashtra, and in the Yakama Nation are just the beginning of a story. The rest of the story is not yet written, but it is about expanding our love of nature beyond an appreciation of its beauty. It is about loving it with our intellects, too, as we come to recognize the ways in which we vitally interact with so many other species, each of which interacts with its own complement of plants, animals, and microbes, in an Earth-sized web of interdependence.

We cannot make it alone. And as soon as we see that, may we fight like hell to defend and nurture the ecosystems native to where we live—whatever combination of forest, wetland, grassland, seagrass, or mangrove that may be—to shelter our houses, farms, and livelihoods as best we can from the sea-level rise, heat waves, floods, drought, and megastorms that are growing only stronger. In nature, we have all the allies we need in this struggle, if only we choose to acknowledge them as such.

ACKNOWLEDGMENTS

I am grateful to all the people featured in this book who shared their forest-planting stories with me in multiple internet-assisted exchanges, and sometimes in person, and to Elise Van Middelem for pointing me toward some of these great leaders. In addition, Enrico Fusto, Damien Saracini, Daan Bleichrodt, Gaurav Gurjar, Kazue Fujiwara, and Elgene Box helped me to understand the subtleties of the Miyawaki Method well enough to attempt to explain it in writing and to troubleshoot the planning process for the mini-forest project in Roscoff. I am also grateful to the people of Roscoff, to the association Ekorrigans, and especially to Aurore Clément for partnering with me in our discovery of the Miyawaki Method and stewarding a project together.

I am grateful to my colleagues at Biodiversity for a Livable Climate for opening the door to the world of ecosystem restoration, and for giving me a context in which to learn about the amazing and amazingly overlooked power of ecology to regulate the climate and keep people healthy and safe. I'm grateful to Fern Bradley and Natalie Wallace of Chelsea Green for guiding me through this book project from conception to completion.

I am grateful to the many people who reviewed sections of the book and offered constructive feedback: Jo Green, Claudia Prado, Philip Bogdanoff, Adam Sacks, Benoit Sarels, Sabine Kerizin, Bianca Virondaud, Mohsen Kayal, and Jane Ballard. I appreciate the scientific input offered to me by Agnès Lieurade and Marc-André Selosse. I am especially grateful to Tomoko Ogawa, who interpreted interviews with translated material from Japanese to English, and reviewed a chapter.

Finally, I am grateful to my family—my parents, Holly and Barrs Lewis, my brother and sister-in-law, Sage and Aleigh Lewis, and my partner, Ehsan Kayal, for being enthusiastic and encouraging throughout the process and for offering helpful feedback on multiple drafts of all the chapters. And I thank my children, Rose and Anahita Kayal, for installing their desks next to my desk to keep me company while I wrote.

RESOURCES

There is a substantial collection of resources available for those seeking to implement the Miyawaki Method or learn more about the theory behind it. The following is a brief selection of useful organizations, books, and articles.

Organizations

Afforestt
www.afforestt.com
Shubhendu Sharma's company, based in India, provides online training courses and an 11-part online video tutorial series (in English) at https://youtube.com/playlist?list=PLDw6Om GaV5rnOCATIho19IcvpF2eqsG_6.

Boomforest
www.boomforest.org/fr
A French organization whose website includes downloadable one-pagers (in French), including an outline of key steps to planning and planting a mini-forest, a sample budget, a list of technical details, and a planting mound diagram.

Earthwatch Europe
earthwatch.org.uk/get-involved/tiny-forest
A UK organization leading a Tiny Forest program based on the IVN Education model.

Ecological Balance
www.ecobalances.org
Limbi Blessing Tata's mini-forest-planting organization in Cameroon.

International Association for Restoration of Native Forest (ReNaFo)
www.renafo.com
A Japanese organization promoting the Miyawaki Method, which authored the publication "Guidelines for Natural Forest Restoration Techniques Using Potential Natural Vegetation" (in Japanese).

IVN Nature Education Tiny Forest
www.ivn.nl/tinyforest
A Dutch organization, whose Tiny Forest program is headed by Daan Bleichrodt. The website has a downloadable handbook called "Tiny Forest Planting Method Handbook" and a short online course (both in English).

MiniBigForest
https://www.minibigforest.com
A French mini-forest-planting organization with multiple project profiles on its website.

Natural Urban Forests
https://www.naturalurbanforests.com
Ethan Bryson's mini-forest-planting company based in Seattle, Washington.

Toulouse in Transition
www.toulouse.entransition.fr/micro-foret-urbaine

A French organization that has a downloadable step-by-step guide to planting a mini-forest (in French).

Urban Forests
http://urban-forests.com/en
A Belgian mini-forest-planting company, with lots of project profiles and other resources online.

SUGi
www.sugiproject.com
A company that provides funding and coordination for mini-forest projects.

Minnesota Department of Natural Resources
www.dnr.state.mn.us/eco/mcbs/vegetation_sampling.html
Has a downloadable field guide for conducting a vegetation survey using the relevé method, entitled "Handbook for Collecting Vegetation Plot Data in Minnesota: The Relevé Method, 2nd Edition."

Books and Articles

The Healing Power of Forests by Akira Miyawaki and Elgene O. Box (Kosei Publishing Company, 2007)

"A Philosophical Basis for Restoring Ecologically Functioning Urban Forests: Current Methods and Results," by Akira Miyawaki, in *Ecology, Planning, and Management of Urban Forests*, edited by Bruce N. Anderson, Robert W. Howarth, and Lawrence R. Walker (Springer, 2007)

Let's Create Indigenous Forest! Continuing for Millennia, from Japan to the World, edited by Kazue Fujiwara (Fujiwara Shoten Publisher, January 2022) (in Japanese) (2022)

"Creative Ecology: Restoration of Native Forests by Native Trees" by Akira Miyawaki (*Plant Biotechnology* 16, no. 1, 1999)

NOTES

Introduction: Restoration in Roscoff

1. United Nations, "Plastic Pollution Set to Double by 2030," accessed January 18, 2022, https://news.un.org/en/story/2021/10/1103692.

Chapter 1: The Miyawaki Method

1. Personal communication with the author.
2. Facebook, accessed September 15, 2021, https://www.facebook.com/watch/?ref=search&v=927185797402261&external_log_id=e675c7d8-af8a-4f31-b5a3-06f38b6d4cea&q=asi+tree+planting.
3. Akira Miyawaki and Elgene Owen Box, *The Healing Power of Forests: The Philosophy behind Restoring Earth's Balance with Native Trees* (Tokyo: Kosei Publishing, 2007).
4. James E. M. Watson et al., "The Exceptional Value of Intact Forest Ecosystems," *Nature Ecology & Evolution* (2018), 1.
5. Miyawaki and Box, *Healing Power of Forests*, 114.
6. Akira Miyawaki, "A Philosophical Basis for Restoring Ecologically Functioning Urban Forests: Current Methods and Results," in *Ecology, Planning, and Management of Urban Forests*, edited by Margaret M. Carreiro, Yong-Chang Song, and Jiangou Wu (New York: Springer, 2008), 188.
7. Suzanne Simard, "How Trees Talk to Each Other," TEDSummit (June 2016), accessed December 14, 2021, https://www.ted.com/talks/suzanne_simard_how_trees_talk_to_each_other?language=en#t-36889.

8. Miyawaki and Box, *Healing Power of Forests*, 254.
9. Miyawaki and Box, *Healing Power of Forests*, 122.
10. Miyawaki and Box, *Healing Power of Forests*, 118.
11. Miyawaki and Box, *Healing Power of Forests*, 126.
12. Akira Miyawaki, "Japanese and Chinju-No-Mori: Tsunami-Protecting Forest after the Great East Japan Earthquake 2011," *Phytocoenologia* (2014), 10.
13. Shubhendu Sharma, "An Engineer's Vision for Tiny Forests, Everywhere," TED (March 2014), accessed December 14, 2021, https://www.ted.com/talks/shubhendu_sharma_an_engineer_s_vision_for_tiny_forests_everywhere.
14. Shubhendu Sharma, "Stop Making Lawns, Plant a Forest!" INKtalks 2012, https://www.youtube.com/watch?v=LrWj1n7lkIs&t=27s.
15. Miyawaki and Box, *Healing Power of Forests*, 214.
16. Miyawaki and Box, *Healing Power of Forests*, 216.
17. "Prof. Akira Miyawaki, Pioneer of Restoration of Native Forests by Native Trees, Learning from Chinjyu No Mori—Sacred Shrine Forests (Part 6)," UBrainTV, United Brain Networks Ltd., 2012, accessed September 15, 2021, http://www.ubraintv.com/watch.php?id=310.

Chapter 2: On the Trail of Ancient Forests

1. Miyawaki and Box, *Healing Power of Forests*, 61.
2. Akira Miyawaki, "Aiming for the Restoration of a Green Global Environment: Restoration of the Green Environment on the Basis of Field Studies and Research into the Ecology of Vegetation," The Winners of the Blue Planet Prize 2006, 5–6, https://www.af-info.or.jp/better_future/pdf/vol_III/2006_winners.pdf.
3. Miyawaki, "Aiming for the Restoration," 6.
4. Miyawaki, "Aiming for the Restoration," 7.
5. Miyawaki, "Aiming for the Restoration," 8.
6. Miyawaki and Box, *Healing Power of Forests*, 61.
7. Miyawaki and Box, *Healing Power of Forests*, 61.

8. Akira Miyawaki, *Creating a Forest of Life* [Akira Miyawaki's autobiography, in Japanese only] (Tokyo: -Fujiwara-Shoten, 2019), 149.
9. Miyawaki, "Aiming for the Restoration," 11.
10. Miyawaki, "Aiming for the Restoration," 11.
11. Miyawaki and Box, *Healing Power of Forests*, 21.
12. Alexis de Tocqueville, *A Fortnight in the Wilderness* (Delray Beach: Levenger Press, 2003), 38.
13. M. D. S. Chandran, M. Gadgil, and J. D. Hughes, "Sacred Groves of the Western Ghats of India," in *Conserving the Sacred: For Biodiversity Management*, edited by P. S. Ramakrishnan, K. G. Saxena, and U. M. Chandrashekara (Enfield: Science Publishers, 1998).
14. Alison A. Ormsby and Shonil A. Bhagwat, "Sacred Forests of India: A Strong Tradition of Community-Based Natural Resource Management," *Environmental Conservation* 37, no. 3 (August 2010).
15. This was not a Miyawaki Method project, although it was similar in terms of the use of multiple native species.
16. "About Us," Kerala Forest Department, accessed September 21, 2021, https://forest.kerala.gov.in/index.php/about-us /introduction.
17. Ormsby and Bhagwat, "Sacred Forests of India," 322.

Chapter 3: Woods and Water

1. David Ellison et al., "Trees, Forests and Water: Cool Insights for a Hot World," *Global Environmental Change* 43 (2017), 57, https://doi.org/10.1016/j.gloenvcha.2017.01.002.
2. A. S. Arya Ajai, P. S. Dhinwa, S. K.Pathan, and K. Ganesh Raj, "Desertification/Land Degradation Status Mapping of India," *Current Science* 97 (November 2009), 1478–83, accessed September 15, 2021, http://www.indiaenvironmentportal.org .in/files/Desertification_0.pdf.
3. Indira Gandhi Institute of Development Research, "Chapter 6: Land Resources and Degradation," *State of the Environment*

Report: Maharashtra 2005–2006, accessed September 15, 2021, http://mahenvis.nic.in/soer_soer5-6.aspx.

4. Maharogi Sewa Samiti, Warora (nonprofit organization), questionnaire by United Nations Development Program, "Become a Biodiversity Champion," Anandwan, Chandapur District, Maharashtra, 10.
5. Maharogi Sewa Samiti, "Become a Biodiversity Champion," 53.
6. Kaveh Madani, "Water Management in Iran: What Is Causing the Looming Crisis?" *Journal of Environmental Studies and Sciences* (August 2014), 6.
7. G. Courade, "The Urban Development of Buea: An Essay in Social Geography," transl. J. P. Magouet (Youande, Cameroon: Orstom, 1972).
8. Roy Lyonga Mbua, "Water Supply in Buea, Cameroon: Analysis and the Possibility of Rainwater Harvesting to Stabilize the Water Demand" (PhD diss., Brandenburg University of Technology, 2013), 45, https://opus4.kobv.de/opus4-btu/frontdoor/index/index/docId/2855.
9. Secretariat of the Convention on Biological Diversity and Central African Forests Commission, "Biodiversity and Forest Management in the Congo Basin" (Montreal: 2009), https://www.cbd.int/doc/books/2009/B-03188.pdf.
10. UNESCO World Heritage, "Natural World Heritage in the Congo Basin," World Heritage Centre, accessed September 25, 2021, https://whc.unesco.org/en/conservation-congo-basin/.
11. Limbi Blessing Tata, "The World Is in a Crisis, Fire, Floods, Typhoons and Hurricanes Everywhere," Kanthari TALKS 2018, https://www.youtube.com/watch?v=mQkhOOW49ow.

Chapter 4: Urban Oases

1. Christophe Najdovski, email communication with author.
2. Marie-Dominique Aponte, "Métier Portrait: Agent de Maîtrise Sylvicole" [Career Portrait: Silviculture Expert], *Présence de l'enseignement agricole privé* 185 (January–February 2008).

3. Meteo France, "Retour sur la canicule historique et meurtrière d'août 2003," Toutes les actualités de Météo-France sur la météo et le climat, en France et dans le monde, accessed September 18, 2021, http://www.meteofrance.fr/climat-passe-et-futur/evenements-remarquables/retour-sur-la-canicule-daot-2003-; Mairie de Paris, *Stratégie de Résilience de Paris*, October 2017, Paris, France, 25, https://cdn.paris.fr/paris/2019/07/24/ebc807dec56112639d506469b3b67421.pdf; Agence Parisienne du Climat, "À Paris comme ailleurs, l'été 2019 a-t-il battu tous les records de chaleur?" September 18, 2019, https://www.apc-paris.com/actualite/a-paris-comme-ailleurs-lete-2019-a-t-il-battu-tous-records-chaleur.
4. Christophe Najdovski, email response to author, June 25, 2021.
5. Agence Parisienne du Climat, "Le changement climatique à Paris: Évolution du climat à Paris depuis 1900, quel climat futur?" July 2015, 6, https://www.apc-paris.com/system/files/file_fields/2015/07/28/plaquetteccaparis-pagesdoublesjuillet2015.pdf.
6. Mairie de Paris, *Stratégie de Résilience*, 28.
7. Intergovernmental Panel on Climate Change (IPCC), "Climate Change Widespread, Rapid, and Intensifying," August 9, 2021, accessed September 18, 2021, https://www.ipcc.ch/2021/08/09/ar6-wg1-20210809-pr/.
8. A. J. McElrone, B. Choat, G. A. Gambetta, and C. R. Brodersen, "Water Uptake and Transport in Vascular Plants," *Nature Education Knowledge* 4, no. 5 (2013), 6.
9. David Ellison et al., "Trees, Forests and Water: Cool Insights for a Hot World," *Global Environmental Change* 43 (2017), 54, https://doi.org/10.1016/j.gloenvcha.2017.01.002.
10. Agence Parisienne du Climat, "L'ilot de chaleur urbain," September 2018, https://www.apc-paris.com/system/files/file_fields/2018/11/07/icu-brochureapc-mf.pdf.
11. National Oceanic and Atmospheric Administration, "Hot Days in the City? It's All about Location," October 15, 2018, https://www.noaa.gov/news/hot-days-in-city-it-s-all-about-location.

12. Cynthia Rosenzweig et al., "Mitigating New York City's Heat Island with Urban Forestry, Living Roofs, and Light Surfaces," A Report to the New York State Energy Research and Development Authority (2006), 1, http://usclimateandhealth alliance.org/post_resource/mitigating-new-york-citys-heat -island-with-urban-forestry-living-roofs-and-light-surfaces/.
13. MIT Sensible City Lab, Treepedia, "Exploring the Green Canopy in Cities around the World," accessed January 18, 2022, http://senseable.mit.edu/treepedia/.
14. Ville de Paris, "Le Plan Biodiversité 2018–2024 pour Paris," April 2019, https://www.paris.fr/pages/un-nouveau-plan -biodiversite-pour-paris-5594.
15. Mairie de Paris, "Guide des plantes natives du bassin Parisien: produites par la ville de Paris" [Guide to Plants Native to the Paris Basin: Produced by the City of Paris], October 2013, https://cdn.paris.fr/paris/2019/07/24/a0ce6796629b45360b8965 aee953922b.pdf.
16. T. Sitzia, A. Cierjacks, D. de Rigo, and G. Caudullo, "Robinia Pseudoacacia in Europe: Distribution, Habitat, Usage and Threats," in *European Atlas of Forest Tree Species*, eds. J. San-Miguel-Ayanz, D. de Rigo, G. Caudullo, T. Houston Durrant, and A. Mauri (Luxembourg: European Commission, 2016), 166.
17. Ville de Paris, "Plan Arbre: Les actions de Paris pour l'arbre et la nature en ville" [Tree Plan: Paris Actions for Trees and Nature in the City] (2021), 18, https://www.paris.fr/pages /l-arbre-a-paris-199.
18. Kersten Nabielek, David Hamers, and David Evers, "Cities in the Netherlands: Facts and Figures on Cities and Urban Areas," PBL Netherlands Environmental Assessment Agency (The Hague: PBL Publishers, 2016), https://www.pbl.nl/sites/default/files /downloads/PBL-2016-Cities-in-the-Netherlands-2470.pdf.
19. Cryuff Foundation, "Cryuff Courts," accessed November 29, 2021, https://www.cruyff-foundation.org/en/activities /cruyff-courts/.

20. "Zaandam," Wikipedia, Wikimedia Foundation, accessed May 3, 2021, https://en.wikipedia.org/wiki/Zaandam.
21. An earlier experimental Miyawaki forest was planted in Sardinia, Italy, in 1997.
22. "What Are the Effects of a Tiny Forest?" IVN Nature Education, accessed September 25, 2021, https://www.ivn.nl/tinyforest/tiny-forest-worldwide/effects-tiny-forest.
23. Caspar A. Hallman et al., "More than 75 Percent Decline over 27 Years in Total Flying Insect Biomass in Protected Areas," *PLoS ONE* 12, no. 10 (October 2017), https://journals.plos.org/plosone/article?id=10.1371/journal.pone.0185809.
24. Francisco Sánchez-Bayo and Kris A. G. Wyckhuys, "Worldwide Decline of the Entomofauna: A Review of Drivers," *Biological Conservation* 232 (2019), https://doi.org/10.1016/j.biocon.2019.01.020.
25. Agence France Presse, "'Catastrophe' as France's Bird Population Collapses Due to Pesticides," *The Guardian*, March 20, 2018, https://www.theguardian.com/world/2018/mar/21/catastrophe-as-frances-bird-population-collapses-due-to-pesticides.
26. BirdLife International, *European Red List of Birds* (Luxemburg: Office for Official Publications of the European Communities, 2015), https://portals.iucn.org/library/sites/library/files/documents/RL-4-020.pdf.
27. Elizabeth Pennisi, "Three Billion North American Birds Have Vanished since 1970, Surveys Show," *Science*, September 19, 2019, https://www.sciencemag.org/news/2019/09/three-billion-north-american-birds-have-vanished-1970-surveys-show.
28. Erik Aerts et al., "Information on LULUCF Actions, The Netherlands," Dutch Ministry of Agriculture, Nature, and Food Quality (2020), 13, https://english.rvo.nl/sites/default/files/2020/12/NLD%202020_529-art10_20201203.pdf.

Chapter 5: Campuses, Coastlines, and Foothills

1. Miyawaki and Box, *Healing Power of Forests*, 155.

2. "ASI Tree Planting," Facebook, accessed November 29, 2021, https://www.facebook.com/ASI-Tree-Planting-518153724972139.
3. Miyawaki, "Aiming for the Restoration," 268.
4. Sango Co., Ltd., Sango Group Report 2018, 5, https://www.sango.jp/en/pdf/sango_groupreport2018eng_all.pdf.
5. Miyawaki and Box, *Healing Power of Forests*, 171.
6. Miyawaki and Box, *Healing Power of Forests*, 169.
7. "Prof. Akira Miyawaki, Pioneer of Restoration of Native Forests by Native Trees, Great Wall of China Forest Restoration Project (Part 5)," UBrainTV, United Brain Networks Ltd., 2012, accessed September 15, 2021, http://www.ubraintv.com/watch.php?id=309.
8. UNESCO World Heritage, "The Great Wall," UNESCO World Heritage Centre, accessed September 25, 2021, https://whc.unesco.org/en/list/438/.
9. International Union for Conservation of Nature (IUCN), "Planning for Up-Scaled FLR in the Miyun Region: Improving Beijing's Water Security through Forest Landscape Restoration," accessed September 25, 2021, https://www.iucn.org/downloads/know_for_flr_case___miyun_watershed_1.pdf.
10. "Prof. Akira Miyawaki, Pioneer of Restoration (Part 5)."
11. Jianguo Liu et al., "Ecological and Socioeconomic Effects of China's Policies for Ecosystem Services," *PNAS* 105, no. 28 (July 2008).
12. Akira Miyawaki, "Creative Ecology: Restoration of Native Forests by Native Trees," *Plant Biotechnology* 16, no. 1 (1999), 23, https://www.jstage.jst.go.jp/article/plantbiotechnology1997/16/1/16_1_15/_pdf.
13. "Prof. Miyawaki, Pioneer of Restoration (Part 5)."
14. Kazue Fujiwara (personal communication).

Chapter 6: Healing Forests

1. Miyawaki, "Aiming for the Restoration," 275.
2. Maryam Nazzal and Samer Chinder, "Lebanon Cities' Public Spaces," *International Journal for Crime, Justice and Social*

Democracy 3, no. 1, https://www.crimejusticejournal.com/article/view/323/0.

3. Adib Dada, "Beirut's RiverLESS Forest," *Society for Ecological Restoration* (October 1, 2020), https://www.ser.org/news/528891/Beiruts-RiverLESS-Forest.htm.
4. UN News, "Beirut Facing Acute Environmental Crisis, Warns UN Energy Specialist," United Nations, September 1, 2020, https://news.un.org/en/story/2020/09/1071462.
5. Aidan Farrow, Kathryn A. Miller, and Lauri Myllyvirta, *Toxic Air: The Price of Fossil Fuels*, Greenpeace Middle East and North Africa (June 2020), 13, https://www.greenpeace.org/mena/en/appr/.
6. Nafisa Eltahir, "UNICEF Warns Millions of Lebanese Face Water Shortages," *Reuters* (August 21, 2021), https://www.reuters.com/world/middle-east/unicef-warns-millions-lebanese-face-water-shortages-2021-08-21/.
7. "Healing Forest: Yakama Nation, USA," SUGi, accessed online December 3, 2021, https://www.sugiproject.com/projects/healing-forest.
8. "Native American Children: Taken from Their Families," PBS Learning Media, accessed December 15, 2021, https://www.pbslearningmedia.org/resource/arct.socst.ush.wounded12aschoolsa/taken-from-their-families/.
9. "Healing Forest: Yakama Nation, USA," SUGi.
10. "Coronavirus (COVID-19) Daily Update," London Datastore, accessed September 26, 2021, https://data.london.gov.uk/dataset/coronavirus--covid-19--cases; Rob England, "Coronavirus: What's the Infection Rate in London?" *BBC News* (January 6, 2021), https://www.bbc.com/news/uk-england-london-55558164.
11. Benoit Faucon and Max Colchester, "'Covid Triangle' Emerges in London's Poorer Neighborhoods as U.K. Variant Rampages," *Wall Street Journal* (January 18, 2021), https://www.wsj.com/articles/covid-triangle-emerges-in-londons-poorer-neighborhoods-as-u-k-variant-rampages-11610968605.

12. "Overview of London Boroughs," Trust for London, accessed September 26, 2021, https://www.trustforlondon.org.uk/data/boroughs/overview-of-london-boroughs/.

Chapter 7: Earth's Living Tissue

1. Personal communication with the author.
2. "Legend of the Forest: Akira Miyawaki, 87, Plants Trees with Children," YouTube, 2015, https://www.youtube.com/watch?v=cfZTzsQ4gEs.
3. Miyawaki and Box, *Healing Power of Forests*, 237; Meghan L. Alovio et al., "Demystifying Dominant Species," *New Phytologist* 223 (2019).
4. Miyawaki and Box, *Healing Power of Forests*, 65.
5. Frank N. Egerton, "History of Ecological Sciences, Part 47: Ernst Haeckel's Ecology," *Bulletin of the Ecological Society of America*, July 2013, https://esajournals.onlinelibrary.wiley.com/doi/epdf/10.1890/0012-9623-94.3.222.
6. Bradley J. Cardinale, "Biodiversity Improves Water Quality through Niche Partitioning," *Nature* 472 (April 2011), 86, https://www.nature.com/articles/nature09904.
7. Cardinale, "Biodiversity Improves Water Quality."
8. Cardinale, "Biodiversity Improves Water Quality," 88.
9. David Tilman, Forest Isbell, and Jane M. Cowles, "Biodiversity and Ecosystem Functioning," *Annual Review of Ecology, Evolution, and Systematics* 45 (2014), 471, https://doi.org/10.1146/annurev-ecolsys-120213-091917.
10. Mar Sobral et al., "Mammal Diversity Influences the Carbon Cycle through Trophic Interactions in the Amazon," *Nature Ecology & Evolution* (2017), https://doi.org/10.1038/s41559-017-0334-0.
11. Sobral, "Mammal Diversity," 3.
12. Sobral, "Mammal Diversity," 4.
13. Dinerstein, "A Global Deal for Nature: Guiding Principles, Milestones, and Targets," *Science Advances* 5 (4) (2019), 3, http://www.science.org/doi/10.1126/sciadv.aaw2869.

14. David U. Hooper et al., "A Global Synthesis Reveals Biodiversity Loss as a Major Driver of Ecosystem Change," *Nature* 486 (2012), 105, https://www.nature.com/articles/nature11118.
15. Sandra Diaz and Marcelo Cabido, "Vive la Difference: Plant Functional Diversity Matters to Ecosystem Processes," *Trends in Ecology and Evolution* 16, no. 11 (2001), 652.
16. M. Loreau et al., "Biodiversity and Ecosystem Functioning: Current Knowledge and Future Challenges," *Science* 294 (2001), 807, https://www.science.org/doi/abs/10.1126/science.1064088.
17. Karin T. Burghardt, Douglas W. Tallamy, Christopher Philips, and Kimberly J. Shropshire, "Non-Native Plants Reduce Abundance, Richness, and Host Specialization in Lepidopteran Communities," *Ecosphere* 1, no. 5 (2010), https://esajournals.onlinelibrary.wiley.com/doi/10.1890/ES10-00032.1.
18. Mark van Kleunen et al., "Global Exchange and Accumulation of Non-Native Plants," *Nature* 525 (2015), https://www.nature.com/articles/nature14910; Julian D. Olden et al., "Ecological and Evolutionary Consequences of Biotic Homogenization," *Trends in Ecology & Evolution* 19, no. 1 (January 2004), https://www.sciencedirect.com/science/article/abs/pii/S016953470300288X.
19. P. Buringh, "Organic Carbon in Soils of the World," in *The Role of Terrestrial Vegetation in the Global Carbon Cycle*, ed. G. M. Woodwell (Hoboken, NJ: John Wiley & Sons, 1984).
20. Rattan Lal, "Beyond COP 21: Potential and Challenges of the '4 per Thousand' Initiative," *Journal of Soil and Water Conservation* 71, no. 1 (2016), 21A, https://doi.org/10.2489/jswc.71.1.20A.
21. Lal, "Beyond COP 21," 21A.
22. Jean-Francois Bastin et al., "The Global Tree Restoration Potential," *Science* 365, no. 6448 (2019), https://www.science.org/lookup/doi/10.1126/science.aax0848.
23. The study specified that this amount of land (1.8 billion ha) could accommodate 0.9 billion ha of canopy. (Since forests can have up to 100 percent of "canopy cover," which is the

area covered by tree crown vertically projected to the ground, it is possible to distinguish between potential *forest land* and potential *canopy cover*.)

24. Thomas Crowther, "The Global Movement to Restore Nature's Biodiversity," TED (October 2020), accessed September 26, 2021, https://www.ted.com/talks/thomas_crowther_the_global_movement_to_restore_nature_s_biodiversity?language=en.
25. European Commission, "The 3 Billion Tree Pledge for 2030" (commission staff working document, Brussels, July 16, 2021), https://ec.europa.eu/environment/pdf/forests/swd_3bn_trees.pdf.
26. Crowther, "The Global Movement."
27. Crowther, "The Global Movement."
28. FAO and UNEP, "The State of the World's Forests 2020: Forests, Biodiversity and People" (Rome, 2020), 15, https://doi.org/10.4060/ca8642en.
29. Kim Naudts et al., "Europe's Forest Management Did Not Mitigate Climate Warming," *Science* 351, no. 6273 (February 2016), 599, https://www.science.org/doi/abs/10.1126/science.aad7270.
30. Food and Agriculture Organization of the United Nations, "Forest Product Statistics," accessed September 26, 2021, http://www.fao.org/forestry/statistics/80938@180723/en/.
31. Enviva, "Displace Coal. Grow More Trees. Fight Climate Change," May 19, 2021, https://www.envivabiomass.com/.
32. Sam L. Davis, "Industry Impacts on US Forests: The Great American Stand Series," Dogwood Alliance, 2018, accessed September 26, 2021, https://www.dogwoodalliance.org/wp-content/uploads/2018/08/Industry-Impacts-on-US-Forests.pdf.
33. "U.S. Forest Resource Facts and Historical Trends," eds. Sonja N. Oswalt and W. Brad Smith, US Forest Service FS-1036 (August 2014), 7, https://www.fia.fs.fed.us/library/brochures/docs/2012/ForestFacts_1952-2012_Metric.pdf.
34. "U.S. Forest Resource Facts and Historical Trends," 8; Natural Resources Defense Council, "In the U.S. Southeast, Natural

Forests Are Being Felled to Send Fuel Overseas" (2015): 9, https://www.nrdc.org/sites/default/files/southeast-biomass-exports-report.pdf.

35. N. L. Stephenson et al., "Rate of Tree Carbon Accumulation Increases Continuously with Tree Size," *Nature* 507 (2014), 91, https://doi.org/10.1038/nature12914.
36. Sam L. Davis, "Treasures of the South: The True Value of Wetland Forests," Dogwood Alliance, accessed January 17, 2022, https://www.dogwoodalliance.org/about-us/forests-of-the-south/.
37. Per Sandström et al., "On the Decline of Ground Lichen Forests in the Swedish Boreal Landscape: Implications for Reindeer Husbandry and Sustainable Forest Management," *Ambio* 45 (2016), https://doi.org/10.1007/s13280-015-0759-0.
38. Anuj Kumar et al., "Forest Biomass Availability and Utilization Potential in Sweden: A Review," *Waste and Biomass Valorization* 12 (2021), https://doi.org/10.1007/s12649-020-00947-0.
39. European Commission, "Renewable Energy—Recast to 2030 (RED II)," accessed December 13, 2021, https://ec.europa.eu/jrc/en/jec/renewable-energy-recast-2030-red-ii.
40. Anton Foley et al., Open letter to President Charles Michel, European Council (March 21, 2021), https://forestdefenders.eu/wp-content/uploads/2021/03/Final-version.open-letter_-on-the-international-day-of-forests.pdf.
41. Anton Foley et al., Open letter to President Charles Michel.
42. Baptiste Morizot, *Raviver les braises du vivant* (Actes Sud/Wildproject, 2020).
43. Morizot, *Raviver*, 101.
44. Morizot, *Raviver*, 106.
45. Starre Vartan, "Beavers on the Coast Are Helping Salmon Bounce Back. Here's How," *National Geographic*, August 13, 2019, https://www.nationalgeographic.com/animals/article/coastal-beavers-help-salmon-recovery-washington.
46. Miyawaki and Box, *Healing Power of Forests*, 59.

47. "What Are the Effects of a Tiny Forest?" IVN Nature Education.

Chapter 8: Mini-Forest Field Guide

1. Miyawaki, "Creative Ecology," 15.
2. Miyawaki and Box, *Healing Power of Forests*, 158.
3. Miyawaki, *Philosophical Basis*, 190.
4. Miyawaki and Box, *Healing Power of Forests*, 159.
5. Miyawaki and Box, *Healing Power of Forests*, 159.
6. Elgene Box in email message to author (September 2021).
7. For example: "The European Beech is exceptionally adaptable to different climatic, geographical, and physical conditions. It is a very competitive species and asserts itself almost everywhere: from rich calcareous to nutrient-poor sandy soils, from mountains to lowlands, and from humid to dry conditions." ("World Heritage Natural Beech Forests," UNESCO World Heritage, accessed September 27, 2021, https://www.europeanbeechforests.org/.)
8. "Plant Communities," Pennsylvania Department of Conservation & Natural Resources, accessed September 26, 2021, https://www.dcnr.pa.gov/Conservation/WildPlants/PlantCommunities/Pages/default.aspx.
9. Miyawaki, "Aiming for the Restoration," 8.
10. Miyawaki and Box, *Healing Power of Forests*, 243.
11. "Resources for Vegetation Sampling," MN Dept. of Natural Resources, accessed September 26, 2021, https://www.dnr.state.mn.us/eco/mcbs/vegetation_sampling.html?__cf_chl_captcha_tk__=pmd_yhNBQn.jp9WCVF8ZaPSPfzXeLmdrTVMDObvB1VYHLyM-1631784415-0-gqNtZGzNAxCjcnBszQh9.
12. Marten Bruns et al. "Tiny Forest Planting Method Handbook," IVN Nature Education (2019), https://www.ivn.nl/tinyforest/tiny-forest-worldwide/resources-and-downloads.
13. "Miyawaki's Rapid Afforestation," Afforest (PDF not available online).
14. "Guidelines for Natural Forest Restoration Techniques Using Potential Natural Vegetation" [in Japanese], NPO International

Association for Restoration of Native Forest (ReNaFo) (Nagano City: Hoozuki Shoseki, Inc., August 2021).

15. Miyawaki, "Philosophical Basis," 191.

Conclusion: Eco-Restoration Starts at Home

1. "The Boomforest Manifesto," accessed December 13, 2021, https://boomforest.org/fr/pages/manifesto.
2. Miyawaki and Box, *Healing Power of Forests*, 167.
3. Frédéric Bioret and Sébastien Gallet, "Caractérisation phytosociologique des chênaies littorales du Finistère," *Revue Forestière Français* 3–4 (2010), http://documents.irevues.inist.fr/handle/2042/38938.
4. Agnes Lieurade, email message to the author, December 7, 2020.
5. Miyawaki and Box, *Healing Power of Forests*, 167.
6. UICN France, MNHN, LPO, SEOF, and ONCFS, "La Liste rouge des espèces menacées en France: Oiseaux de France métropolitaine" [The Red List of Threatened Species in France: Birds of Metropolitan France] (Paris, 2016): 31, https://inpn.mnhn.fr/docs/LR_FCE/UICN-LR-Oiseaux-diffusion.pdf.
7. Intergovernmental Panel on Climate Change (IPCC), "Summary for Policymakers," *Climate Change 2021: The Physical Science Basis. Contribution of Working Group I to the Sixth Assessment Report of the Intergovernmental Panel on Climate Change*, eds. V. Masson-Delmotte et al., https://www.ipcc.ch/report/ar6/wg1/.

INDEX

Note: page numbers preceded by *ci* refer to images in the color insert.

ABOUT THE AUTHOR

Studio KP

HANNAH LEWIS is the editor of *Compendium of Scientific and Practical Findings Supporting Eco-Restoration to Address Global Warming*, published by Biodiversity for a Livable Climate, a nonprofit environmental organization based in Cambridge, Massachusetts. She has worked in various roles related to building sustainable food systems, including as the Midwest regional director for the National Center for Appropriate Technology. She has an MS in Sustainable Agriculture and Sociology from Iowa State University and a BA in Environmental Studies from Middlebury College. Born and raised in Minneapolis, Minnesota, she lived in France with her partner and their two children during the writing of this book. To learn more, visit: www.hannahlewis.org.